Fundamental Aspects of
ELECTROCRYSTALLIZATION

Fundamental Aspects of
ELECTROCRYSTALLIZATION

J. O'M. Bockris and G. A. Razumney

Electrochemistry Laboratory
John Harrison Laboratory of Chemistry
University of Pennsylvania, Philadelphia, Pennsylvania

PLENUM PRESS · NEW YORK · 1967

ISBN-13: 978-1-4684-0699-3   e-ISBN-13: 978-1-4684-0697-9
DOI: 10.1007/978-1-4684-0697-9

Library of Congress Catalog Card No. 66-22123

Preface

This monograph is an attempt to present in a concise manner those aspects of the electrolytic growth of crystals considered to be of basic mechanistic significance. The accent has been put consistently on the understanding of the physical models of the processes discussed. Out of the extensive amount of information in this field, the authors have tried to abstract those systems which can be considered central, and which can, therefore, be related to the basic concepts connected with the electrodeposition of metals. In particular, stress has been laid upon consideration of atomic movements —of ions and molecules in solution, of adsorbed substances on the surface of metals, of steps on a growing crystal.

Although the main stress is upon the physical picture, a number of derivations have been presented in detail. It is assumed that the reader is familiar to some degree with the concepts of electrodics; in particular, frequent reference is made to models of the electrical double layer at the metal-solution interface and the transfer of charged species across it. This "snapshot" of the fundamentals of electrocrystallization at its present stage of development is mainly directed toward three groups of readers: (i) to the technologist familiar with the many (and theoretically complex!) technological aspects of the deposition of metals who wishes to gain some insight into the fundamentals of the mechanisms occurring; (ii) to the scientist with a general background in physical chemistry who wants to launch himself into the field of fundamental studies on electrocrystallization without devoting excessive time to the task; (iii) to the electrochemist who wishes to deepen his general knowledge of the theory of electrode processes by a more detailed study of the area of electrocrystallization with its particularly interdisciplinary character.

Some comparison of this monograph with the major German work by H. Fischer (1954) is certain to be made. The aims of the present monograph are narrower, and more central, than those of the comprehensive and descriptive presentation of the whole field, as then known, in Fischer's book. The attitude here has been to take those (as yet still simple) aspects of the mechanism of metal deposition that are to some extent now understood in terms of atomic movements. This has, of course, led to several limitations on the areas discussed. Indeed, with the exception of Kardos' theory of leveling and some work on brightening, the mechanistic theory of the formation of polycrystalline deposits is largely a virgin area, due to the virtual lack of knowledge of relative growth rates for the various individual crystal planes which make up the metal surface.

Electroplating, the electrochemical extraction of metals from ores and mixtures, and electrodeposition from nonaqueous solutions, together make up a large area in technology. It is relevant to distinguish two classes of technologies: those in which the fundamentals and theory were understood first and the applications followed in a realized way; and those in which the applied side began as a kind of art and the theory limped behind the art, sometimes dragged on by it and occasionally giving it a little push. Of course, the atomic energy and electronics industries are ideal examples of the first two. The electrodeposition and electroextraction industry is a fair example of the second. It is clear enough that the rate of development of the first type of industry is much greater than that of the second. One can estimate what is possible and at least the direction in which one should push to try to attain it, whereas progress in the second type relied in the past on what might be called "inspired groping." It is hoped that this small monograph may provide a basis for the training of research workers who can then carry out the conversion of what remains an art in technological electro-crystallization into a technology with a largely rational basis.

The authors' special thanks go to the American Electroplaters' Society for their financial support during the writing of the book and to those who worked in the senior author's laboratory on the mechanism of metal deposition. In particular, the authors wish to thank Dr. Otto Kardos for his many erudite comments on and corrections of the manuscript, and Dr. A. Damjanovic, who has taught them much of what they know about the crystallographic aspects of the field. They wish to thank Dr. Dennis Turner, Dr. J. V. Petrocelli, and Dr. R. Weil for many helpful and clarifying comments. The senior author is particularly grateful to Mr. E. Bowerman, whose unrelenting criticisms, withstood for over a decade, have taught him much and often stimulated him to higher standards.

Philadelphia
November, 1966

Contents

Chapter 13

Electrocrystallization and Crystallization from the Gas

Chapter 14

Chapter 1

Retrospect

1. THE FIRST CONTRIBUTIONS

Faraday's laws were enunciated in 1834, and because they were experimentally established[1] by means of the electrodeposition of Sn, Pb, and Sb onto Pt, it is probable that the study of electrocrystallization is one of the oldest parts of experimental electrochemistry. Correspondingly, early concepts of nucleation were mentioned by Gibbs[2] in 1878, and the "discharge" process was discussed by LeBlanc[3] in 1896. The first observations of the form of electrodeposits under a microscope were recorded in 1905 by Huntington,[4] and the first X-ray examinations were made by Glocker and Kaupp in 1924.[5]

In spite of this early beginning, however, the understanding of electrocrystallization in terms of atomic models made very little progress until some ten years ago. The reasons might be said to epitomize those which have retarded the progress of many of the fundamental—and therefore the dependent applied—aspects of electrochemistry (particularly in Electrodics[6]); namely, the obvious possibilities of the practical application of so many parts of the field make the investigations too "applied" at a too early stage, i.e., before the fundamental work has been done, and sufficient modelistic background worked out.* Compared with the fundamental work which

---

* At present, the eagerness to develop practical electrochemical energy generators only some five years after the beginning of widespread research in the field of electrochemical energy conversion (less than 15 years after electrode kinetics became a recognized subject, and before the existence of a text book in English) represents a similar trend.

1

has been carried out during the last decade, that published prior to about 1950 suffered the following drawbacks:

1. It was overwhelmingly occupied with qualitative descriptions of crystal growth forms. The introduction of electron diffraction to such studies by Thompson[7] in 1931 and the Group at Imperial College, London, directed by Finch,[8] with its ability to give information concerning the *crystallographic* properties of deposits of the order of 100 Å in thickness, e.g., in respect of epitaxy, did not aid the study of the atomic processes across the double layer.* Also, suggestions[8] concerning the effect of the strong electrostatic field connected with projecting portions of the growing surface were made (material would preferentially deposit there), whereas no explicit account was taken of the mass transport control of such high (local) current densities.

2. The concept of edge dislocations in metals was suggested by Taylor[10] in 1934, and that of screw dislocations by Burgers[11] in 1939, but it was not until 1949 that it was shown by Burton, Cabrera, and Frank[12,179] that imperfections in the surface of the metal would make unnecessary nucleation as the beginning of crystal growth—at least at low supersaturations (overpotentials).† Hence, attempts at theoretical interpretations of (particularly single) crystal growth before about 1950 laid too much stress upon nucleation kinetics. For example, perhaps the most well-known fundamental contribution to metal deposition kinetics of the 1930's is the treatment of rate-determining three- or two-dimensional nucleation by Erdey-Gruz and Volmer,[14] giving rise, respectively, to the equations

$$\ln i = A - B\eta^{-2}$$

$$\ln i = A' - B'\eta^{-1}$$

* An analogous tendency could still be seen in the well-known book of H. Fischer.[9]

† The formal relationship between the electrodeposition parameters and those pertaining to crystal growth from the vapor was first set out by Mehl and Bockris.[13]

where i is the current density, and η is the overpotential. The realization that defects contribute fundamentally to the structure of a metal surface, and act as sites of growth on the crystals from atoms delivered from vapor or solution,[15] has led investigators to reduce radically the importance which they assigned to nucleation processes.

3. The principal electrodic parameter which controls the rate of electrodeposition is the overpotential, but the concept of this quantity (though correctly used and defined by Erdey-Gruz and Volmer in the fundamental paper[16] of 1930, in which they established the relation between current and potential for charge transfer at an electrode) until the 1950's was a fuzzy concept for most physical chemists, crystallographers, metallurgists, and even electrochemists. Consequently, many publications in the field do not record the overpotential (or even the current density!) at which the observations were made, and thus preclude the possibility of electrodic analysis.

4. The marked effect of trace impurities in the solution on electrode processes at solids was demonstrated for the first time in 1949.[17] Effects on the growth form of electrodeposited materials can be detected when the surface coverage is 10^{-4} and the solution concentration less than 10^{-9} mole/liter.[18] Theories which are offered to explain why the steps which move across the surface in metal deposition are not monatomic, but are several thousand angstroms high, call fundamentally upon the effect of surface adsorbed "inhibitors."[19,20] Publications dated before 1950 were produced by workers who had had little warning of the remarkably low threshhold for the beginning of the effects of trace contaminants.[17] They lacked the techniques[21,22] by which the impurities might have been removed.

5. Corresponding to (3), electrodic (including double-layer) concepts became generally available in book form* only

* The first article in English which attempted to summarize electrode kinetics was published in a book in 1954.[23] Frumkin's "Electrode Processes" was published in 1952. Vetter's "Elektrochemische Kinetik" was published in 1961.[59]

in the 1950's. Until recently, it was difficult for workers in the crystal growth field to find the basis for replacing the simple kinetic theory (gaseous atoms colliding with a surface) with a more elaborate one (diffusion in solution followed by charge transfer at the interface—dependent upon double-layer structure and charge on the metal).

6. Lastly, compared with other electrodic processes, electrocrystallization introduces the fundamental difficulty that the surface changes as the process continues. Basic to approaches to meeting this difficulty is the concept of transient analysis,* but the experimental background thereof involves electronic techniques. String galvanometers were used in experimental electrode process work by Bowden and Rideal[26] and by Baars[27] in 1928. Cathode-ray oscillographs were, however, still rare in 1945, and not commonplace until after 1950.

2. EARLIER CONCEPTS

An observation which was basic to the views of the period 1920–1950 concerning metal deposition was that the number of nuclei on the growing surface decreased with a decrease in current density (Aten and Boerlage, 1920[28]). This stress upon nucleation was a forerunner of the theory which dominated concepts in crystal growth during the 1930's and 40's, namely, that suggested by Kossel,[29] and developed by Stranski.[30] The basic idea is that of two-dimensional nucleation, after which layer growth occurs, the recently deposited atom traveling over the surface planes until it strikes the edge of the advancing layer. It then travels along this edge until it finds the characteristic position of high coordination known as a "kink site" (see Fig. 1). There it stops, and the movement of arrival from the nonmetallic phase, traveling across the surface, colliding with an edge, and finding the kink site is again repeated many times by following atoms.

---

* The first published suggestion in this direction was due to Volmer, in 1933,[24] and the first transient measurements applied to the examination of metal *dissolution* was due to Rojter, Juza, and Polujan, in 1939.[25]

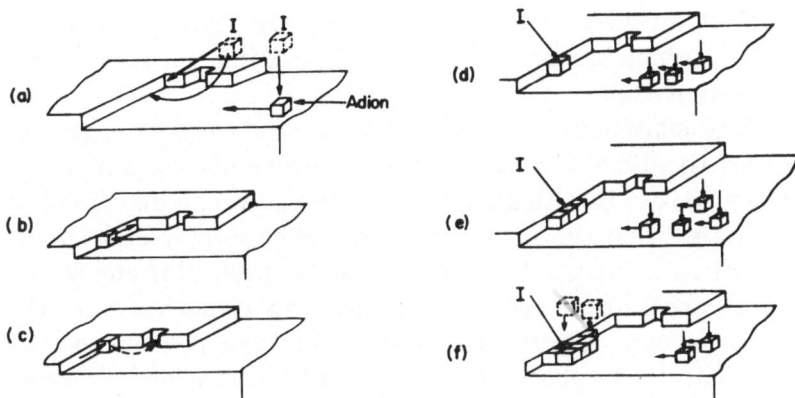

Fig. 1.   Alternative and consecutive steps in crystal building: (a) shows
          direct transfer of an ion I to surface, edge, or kink sites upon
          a metal surface; (b) and (c) show surface and edge diffusion of
          the adion, and in (c) the adion is shown at its kink position;
          (d) shows arrival of other adions, which by surface diffusion build
          around the adion I at the edge; (e) and (f) show incorporation of
          I into the crystal by arrival of other adions migrating across
          the surface or transferred from the solution. Nucleation (not
          shown) between diffusing adions could also occur as a crystal-
          building process (Conway and Bockris[53]).

After building up of one such layer, nucleation must again
occur and a further layer is thus commenced. This layer
covers the atoms which were incorporated in the previous
layer and hence these become part of the metal.

   Kossel and Stranski had been interested in crystal growth
as such. Erdey-Gruz and Volmer utilized their concepts in
furthering the picture of electrolytic crystal growth from
solution.[14] They considered the process of charge transfer in
general, and reported linearity of current density as a function
of potential over quite wide ranges of potential under certain
solution conditions for silver and a number of other metals.
They deduced an equation relating current density to overpotential on the assumption of two-dimensional layer formation
as the rate-determining step (cf. the Kossel–Stranski theory),
and suggested that the metal ion did not undergo charge
transfer at the point at which it arrived in the double layer,

but first diffused *in solution* until it was opposite "a growth site" (presumably an edge or even a kink in the Kossel–Stranski model).

The contributions of Volmer[31] to the study of crystal growth, during the 1930's, were many, and are summarized in a well-known book, "Das Elektrolytische Krystallwachstum." Volmer suggested[24] that the difficulty of the change of metal surface with time in metal deposition could be avoided if one is careful to "choose the experimental condition in such a way that the secondary effects are avoided and the primary electrochemical effects observed." The contributions of Volmer and his school were the most important in the field in the pre-1950 period.

Volmer's work did not, however, go uncontradicted. Thus, Brandes[37] stressed the Kossel–Stranski model for electrolytic crystal growth, i.e., charge transfer from the metal to the ion *at the point of arrival* and diffusion to growth sites, in contrast with Volmer's view that the transfer was directly to a growth site after lateral diffusion in the double layer. Both suggestions were, of course, qualitative.

The other lasting contributions of the pre-1950 period can be divided into three groups, the earliest of which seems to have the most direct connection to the present period, i.e., the work of Kohlschütter,[32] who established much of our early knowledge of nonmorphological phenomenology. Thus, with Torricelli[33] he measured the rate of spreading of layers (approximately 4×10^{-2} cm/min for silver from concentrated $AgNO_3$ solution).* They suggested (but without supporting arguments) that material did indeed reach growth points by means of surface diffusion (i.e., charge transfer occurs to planar sites), and that in layer growth the rate-determining step is the aggregation to a growth site. A particular purpose of their observations was to attempt a qualitative discussion of mechanism on the basis of evidence from current-

---

* However, the calculated local current density from this result is approximately 10 A/cm², a result which seems inconsistent with the conditions of the experiment.

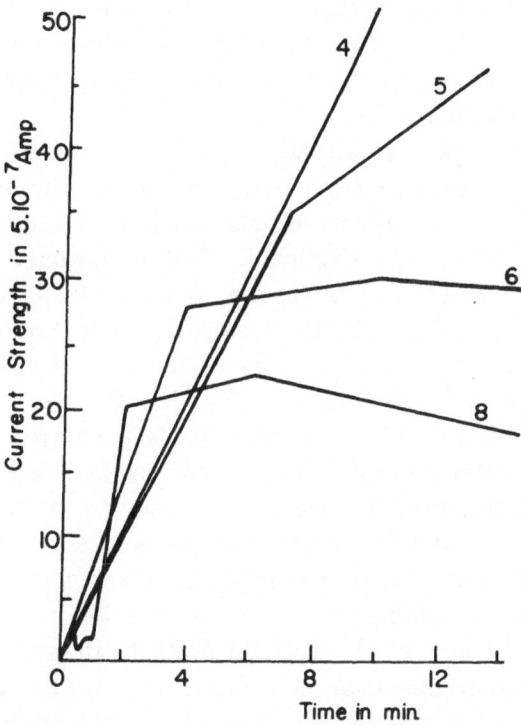

Fig. 2. Current-time curves for Ag deposition (Kohlschütter and Torricelli[33]).

time curves (for times of approximately 10 min). Typical of
these curves is that of Fig. 2. The sudden changes were
explained in terms of "activation and passivation" of the
surface.   Kohlschütter and Torricelli noticed that layers
growing over the metal crystal would remain unchanged and
continue where they had left off, on switching off and on the
depositing current, respectively, but that if the time interval
was too great (3–5 sec) the fresh surface became inactive.
They suggested that this was due to "local action."* They
also made the first suggestion of a "local exhaustion" theory.
There would be a particularly high current density at edges
and corners (high field) where a layer "begins." The solution

* The explanation was inexplicit; but see Chapter 7.

in the region of commencement is exhausted, and the predominant site where charge transfer takes place is now mainly on the planes from which, by diffusion, the moving edges are supplied with atoms.

Fischer's work was already well known in the 1940's and two of his contributions from this period must be mentioned.[34] First, the well-known effects of gelatin addition (e.g., inhibition of dendritic growth) are explicable if it is assumed that substances do not adsorb uniformly but on "active sites."* Metals are very different in their sensitivity to impurities and this was interpreted correctly as a function of the different surface energies.[36] In general, Fe, Ni, Pt, etc. (transition metals) are "hard" (have a high surface energy and easily accept adsorbents) thereby lowering the surface energy. The elements Pb, Sn, and Hg are "soft" (have a relatively low surface energy), and hence the energetics of adsorption is less favorable. The metals in the first group are more susceptible to the effects of inhibitors.

Lastly, Finch and the London School, in spite of their neglect of electrochemical considerations, made some interpretations which are of lasting importance, particularly to our understanding of thicker deposits. Thus, the dependence of the growth process upon the crystal face, also discussed by Kohlschütter and Torricelli, was stressed. The reasons for the breakdown of epitaxy were discussed: it was often due to the inclusion of foreign particles, e.g., co-deposited metals or hydrogen, colloidal particles, etc., in the growing deposit. Amorphous and random deposits were interpreted in terms of adsorbed H of the surface. This H blocks the surface and prevents regular layer growth and hence crystallization. Random deposits arise essentially from the same cause: the adsorbed H is an inhibitor and reduces surface mobility of the diffusing adatoms.

---

* "Active sites" is of course a much used *omnis coverendum* term. No evidence is obtainable from the new work on the dependence of the coverage of electrodes with organic adsorbents as a function of potential,[35,84,125,141] because the sensitivity is too small. But sensitivity of the growth forms to impurities[18] demands heterogeneity of adsorption.

Chapter 2

Perspective

It is necessary, before venturing further, to try to put the various contributions which can be made in metal deposition studies into perspective. The need here is greater than in many fields in electrochemistry because (a) the number of primarily technological publications is very great, so that papers written to elucidate mechanisms may be hardly noticed; (b) it is often difficult to see much relation, say, between an electron microscope study of the form of deposited crystals and the determination of the AC impedance involving the $M^{z+} + ze \rightleftharpoons M_{ads}$ reaction; (c) there is particularly little relation between conceptual and chronological development (i.e., to some extent, fundamental information, for instance, the velocity of movement of a macrostep, was available some 25 years before corresponding information on the velocity of surface diffusion of the products of charge transfer was known).[13,38] Correspondingly, technological development has continued to advance with a maximum of empiricism and perhaps less connection with fundamental research than in many other technological fields.

A curious feature of the basic studies of electrocrystallization is that the dissolution of metals is seldom studied, although in all other branches of electrode kinetics measurements of rates in both directions are carried out.*

---

* General equations which indicated the desirability of doing this have been available since 1951.[39]

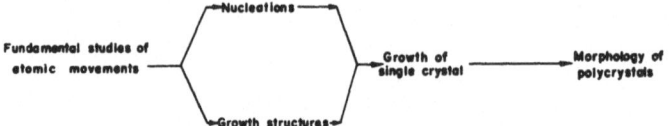

Fig. 3.   Perspectives in studies of metal deposition.

Some perspective can be gathered from Fig. 3, in which the development is represented in a conceptually consecutive way (*cf.* Table 1).   The fundamental approach, i.e., consideration of the surface structure and the movements of the arriving ions on an atomic scale, was begun only in the 1950's.   It led at once to a great improvement in the model for the deposition process of single ions, that is, the charge transfer process and the subsequent diffusion to and incorporation at growth sites.   More recently, the crystal growth process (multilayer deposition) has been analyzed on the same basis, and considerable information has been obtained on the growth mechanism of different crystal forms.[19,20,42,45,47,50]

The theoretical studies of steady-state currents controlled by two-dimensional nucleation (Erdey-Gruz and Volmer[14]) were carried out much earlier than theoretical studies of processes controlled, e.g., by surface diffusion, (see Table 1) but they progressed little because of the absence of experimental verification.   This we can now easily rationalize because we know of the existence of dislocations in metals, which in 1931 was an unrealized structural concept.   A major change in viewpoint came with the introduction, by Burton, Cabrera, and Frank in 1949,[12,44,179] of the concept of emergent screw dislocation into the model for crystal growth.   It was no longer necessary to assume nucleation as a step in the mechanism of growth so long as the concentration of the defects on the surface was not too low or the current density too high. The expected growth spirals were, indeed, soon observed in crystals grown electrolytically,[46] and the young field of the kinetics of electrode processes was applied to express the current at the interface in terms of a model involving spiral growth.[45]

Table 1

Chronology of Outstanding Fundamental Studies in Metal Deposition
1927-1966

Date	Workers	Summary of main results
1927	Kossel;[29] Stranski[30]	Crystal growth occurs in repeated monatomic layers after two-dimensional nucleation.
1931	Erdey-Gruz and Volmer[14]	Equations for current with rate-determining formation of two- or three-dimensional nuclei, discharge, layer spreading.
1932	Kohlschütter and Torricelli[33]	Moving layers observed, travel at approximately 10^{-3} cm/sec; current-time at constant potential is sporadic.
1949	Burton, Cabrera, and Frank[12,44,179]	Crystal growth can occur without preliminary nucleation by means of the rotational movement of a growth spiral originating at a screw dislocation.
1952	Steinberg[46]	Observed growth spirals on electrodeposited Ti.
1954	Lorenz[40]	Consideration of rate-determining surface diffusion of adatoms.
1954	Fischer[9]	Classification of polycrystalline deposits; general application of effect of adsorbed layers on determination of crystal type.
1955	Kaischew and Mutaftschiew[43]	Confirmed Volmer's equations for three-dimensional nucleation.
1956	Vermilyea[45]	Equations for rate of deposition in terms of screw dislocation growth model; "long time transients" discussed qualitatively.
1956	Kardos[168]	Quantitative theory given for action of leveling agents in technological plating.
1957	Mehl and Bockris[13,38]	Deduction of adion concentration from galvanostatic transients.
1957	Vermilyea[58]	Nucleation overpotential experimentally observed for first time (low dislocation density substrate).
1958	Gerischer[41]	Deduction of charge on adion.
1958	Conway and Bockris[52,53]	Energetics of charge transfer ion-metal shows direct deposition to defect sites improbable; adions.
1958	Cabrera and Vermilyea;[20] Frank[19]	The stable existence of macrosteps depends on adsorbed impurities.
1960	Fleischmann and Thirsk[42]	Anodic potentiostatic transients can be applied to give information on mechanism of crystal growth stages.
1963	Mullins and Hirth[170]	Microscopic theory of bunching of microsteps.
1965	Damjanovic, Paunovic, and Bockris[80]	Optical determinations of mechanics of lateral and vertical step movements.

Some 15 to 20 years before the advent of the spiral-growth model, good photographs were available of growing layers (Kohlschütter and Torricelli, 1932). With them arose the considerable problem as to why the moving layers could be seen in an optical microscope, i.e., were at least 2500 Å in height. Elementary attempts to explain this were many[9,48] but it was Frank,[19] and Cabrera and Vermilyea[20] (1958) who simultaneously applied Lighthill and Whitham's theory of kinematic waves[49] to the migration of steps over the surface of crystals. The "wave motion" (cf. Chapter 10, Sections 6 and 7) results from the condition of conservation of the number of microsteps; the shock wave is a result of the interaction between adjacent steps (e.g., by competition for the depositing ions). The "bunch" (or region of high concentration of microsteps) can be expected to develop (or even to be stable) only in the presence of adsorbable impurities (cf. Ref. 170). This was a relatively sophisticated approach and it was soon joined by a theory of whisker[50] and of dendrite[47] growth, each of which facilitated the calculation of the rate of growth numerically with reasonable accuracy—for the first time in the development of electrocrystallization theory.

The papers mentioned are considered the principal ones in the development of the field. They are few in number. In summary, it can be said that a fair body of knowledge is now available concerning happenings to particles on the way to the growth site, and a smattering of knowledge on how very small single crystals grow. Concerning electrocrystallization of macrodeposits, the polycrystals, and the basis of the process of electroplating, a beginning has been made by Fischer[9] in his extensive classificatory and qualitative work. (He has contributed much by his stress upon inhibition as a general phenomenon in electrodeposition[55]). Isolated aspects of the theory (e.g., that of texture[56,57]) are understood; however, ambitious theoretical treatments can only come after a much more advanced theory for the simpler processes is available.

In this monograph, stress is laid upon perspective, and the presentation is intentionally less specialized than that of Bockris and Damjanovic.[51]

Chapter 3

Methods of Investigation

The purpose in any investigation of a metal deposition-dissolution system is to obtain the mechanism of the process

$$(\text{Metal ion})_{\text{solution}} + (z \text{ electrons})_{\text{crystal}} \rightarrow (\text{Metal atom})_{\text{crystal}}$$

"Determination of the mechanism" in the case of a redox reaction or gas evolution at an electrode usually means obtaining the path of the reaction and the rate-determining step in the consecutive sequence.[59] However, in the case of metal deposition two complications arise: (a) the rate constants and intermediate concentrations (and, therefore, the rate-determining step) may change with time; (b) the type of process observed in the crystallization stage is a cooperative one in which the representation of the process in terms of a chemical reaction is difficult.

There are basically two methods which can be applied to the investigation of this problem: (a) electrochemical kinetics, carried out especially in terms of transient experiments because of the probable change with deposition of the kinetic parameters (dependent on the substrate structure) and because the analysis of the complete response of the system to a step function gives more information on the mechanism, in particular concerning intermediates, than the steady-state behavior; (b) surface observational methods, in particular optical methods.

15

1. ELECTROCHEMICAL METHODS

(i) Galvanostatic Transients

This first of the transient methods is applied at a variety of current densities. The limits are often given by (lower limit) the time of charging of the double layer and (upper limit) the transition time. Processes with rate constants up to about 20 cm/sec can be studied.[60] Variables are the normal variables of electrode kinetic examinations, namely, concentration of the reactants, current density, and temperature, together with the structure of the substrate. One should be able to obtain the concentration of intermediates, the heat of activation for diffusion across the surface, the charge on the adion, and the exchange current density.[13,25,41] This method has now been applied extensively.[13,25,38,41,61-68,164]

There is, because of the change of surface with time, no point in "long-time steady-state" measurements, but various "transient steady states" will be encountered (see Chapter 5).

(ii) Potentiostatic Transients

The variables and quantities which may be obtained are essentially the same as those in the galvanostatic method.[41,42,106] However, a more accurate solution can be attained more easily under potentiostatic conditions.

The potentiostatic method has been analyzed in great detail, with special reference to application at long times (i.e., 1–100 sec), by the Newcastle Group under Wynne-Jones, in particular by Fleischmann and Thirsk (who have worked on crystal growth processes analogous to those of metal deposition, e.g., oxide-formation kinetics). They have taken into account the effect of the rotating growth spiral, i.e., the change in surface-diffusion kinetics with time, on the growing surface. Thus, the overpotential dependence of the critical radius for a two-dimensional nucleus results in a time-dependence of the distance between steps originating from a screw dislocation. This time-dependence would impose an upper limit to the time interval in which the distance between steps can be considered constant and, hence, the potentiostatic current-time curve can be interpreted analytically.

Fleischmann and Thirsk[69] have analyzed theoretically the long-time behavior of systems in which growth occurs at discrete centers, preceded by two- or three-dimensional nucleation. The analysis rests upon the assumption that the mechanisms and rate constants for nucleation and growth are independent of each other. The overall rate is then given by the superposition of both processes, and the rate of electrocrystallization will result by summation (or integration) over the individual centers, which appear at the surface according to the nucleation kinetics (assumed in that model to be of 1st order). For each growth center a constant current density* is assumed. Hence, the current per growth center will vary with its age following the variation of surface area (or edge length) and this, in turn, depends on the geometry of the particular case. Thus, if N_τ is the number of nuclei formed per unit area after a time τ ($dN_\tau/d\tau$ is therefore the rate of nucleation), and $j(u)$ is the growth rate of a single isolated nucleus as a function of its age u, the total current density at time t is given by

$$i(t) = zF\int_{\tau=0}^{\tau=t} j(t-\tau)dN_\tau$$
$$= zF\int_{\tau=0}^{\tau=t} j(t-\tau)\frac{dN_\tau}{d\tau}d\tau$$
$$= zF\int_{u=0}^{u=t} j(u)\left(\frac{dN_\tau}{d\tau}\right)_{\tau=t-u}du$$

The preceding model breaks down when the centers grow to such an extent that they start overlapping. The initial geometry is then distorted, and the overall current density will reach a maximum and decrease, or maintain a constant value according to the particular case.

The exact treatment when allowance is made for overlap is difficult without simplifying assumptions regarding the

---

* Or current per unit length of edge.

geometry of the centers, the isotropy and uniformity of the system, etc. Alternatively, a simulation technique can be used in analyzing complicated situations. Independent of the type of analysis employed, the information obtained can be expressed essentially by means of two parameters: the rate constant for nucleation and the rate constant for growth. These kinetic parameters are, in turn, a function of the usual kinetic and structural variables: overpotential, concentrations, temperature, surface tension, etc.

Thus by investigating the potentiostatic behavior of a system over the complete time range from 10^{-6} to 10^3 sec it will be possible to obtain information on the mechanism of the monolayer deposition stage as well as the mechanism of the crystal growth process. This type of approach has hardly as yet been applied to metal deposition,[180] although it has been used for anodic electrocrystallization processes.

Often, it is helpful to use both galvanostatic and potentiostatic transients; their joint use is experimentally very simple, and can be fruitful. For example, it can be shown that a distinction can be made between rate-determining surface diffusion and rate-determining charge transfer simply by comparing the first rise times with each other after the charging process in the galvanostatic and potentiostatic cases, respectively. If $\tau_{galvano}/\tau_{pot} > 2$, surface diffusion is in rate control.[106]

(iii)   Impedance

A basic use of impedance measurements is to attempt to deduce the double layer capacity and thus obtain information concerning roughness (but see below). The adion concentration very near to the reversible potential can be derived from the resistive component of low-amplitude measurements.[70,71] If organic compounds are present, the change of capacity (with respect to the same system in the absence of organic substances) can be related to the surface concentration of the organic compound.[72] Impedance measurements can, then, be employed usefully in inhibitor studies, on the same setup as the other transient measurements.

However, there are two difficulties often overlooked. Vetter[59] pointed out the first, namely, that for solid electrodes the microcrevices contribute to the resistance of the interface to an extent which is frequency dependent because of variations in the length of penetration of the concentration wave in the crevice (which is frequency dependent). Thus, the R vs. $1/\omega^{1/2}$ plot is no longer linear. There is a separate difficulty, as pointed out by Bockris and Conway.[73] Water dipoles are adsorbed upon the surface of an electrode. When an oscillating field is placed upon the electrode, the dipoles orient in a way which at low frequencies follows the applied field. Consequently, they contribute a current within the double layer which depends upon their rotation, giving rise to a fundamentally capacitative "dipole admittance." As the frequency increases and becomes comparable to the relaxation frequency for rotation of the water dipoles, this admittance becomes more resistive. The relaxation time of the dipoles will be thermally distributed so that a certain number will react strongly to a field of any frequency. Thus, the double-layer impedance is fundamentally frequency dependent due to causes not connected with Faradaic processes at the interface.[74] Furthermore, in the case of solid electrodes, that frequency dependence will be affected by the substrate structure, as the heat of adsorption of water will vary with the latter. Hence, one must be cautious in the application of the impedance method to studies in metal deposition, whenever that study involves frequency variation, because it is difficult to extract from impedance measurements the desired kinetic parameters unless the capacitance of the double layer is assumed to be frequency independent and its parallel resistance infinite.

(iv) The Electrochemical Generation of New Surfaces

All work at surfaces suffers from a difficulty of reproducing the "activity" of the surface. Its variation for the first few seconds after the production of a new surface may be quite large (Fig. 4).[68] By means of metal dissolution *in situ*, this factor could be controlled. The difficult problem

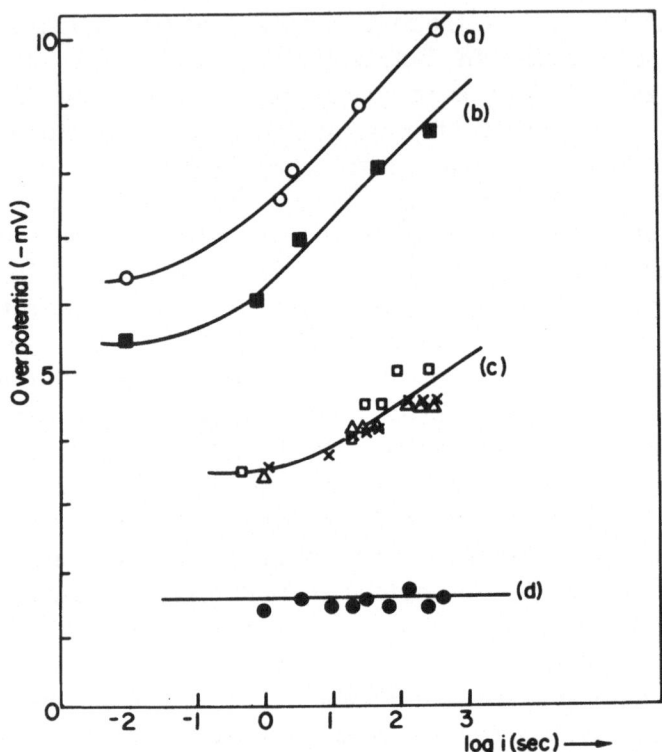

Fig. 4. Change of steady-state overpotential (corresponding to $i = 8.7 \times 10^{-4}$ A/cm²) with time of contact with solution (in the absence of current) after anodic dissolution for Cu electrodes: (a) and (b) He-quenched electrode; (c) electrodeposited electrode; (d) oxide film electrode (Bockris and Kita[68]).

of an irreproducible surface would thus find an easier solution in the electrodic case than in that of gas-phase studies involving surfaces.

2. OPTICAL METHODS

(i) Light Microscopy

There is considerable opportunity for the exploitation of

new microscopic techniques,* in particular Nomarski *polarized interferometry*.[75] The optical system is essentially a differential interferometer. In an ordinary interferometer the surface under observation is compared with a reference (flat) surface, and the optical arrangement is such that the introduction of a small angular deviation between the two beams produces a series of fringes which are actually contour lines for the spacing between specimen and reference plane. In the case of the differential interferometer the reference plane is replaced by a second image of the observed surface. The separation of the beam coming from the object into two slightly deviated components is accomplished by the use of a Wollaston prism, which at the same time introduces a phase shift that varies linearly with distance from the center of the prism (Fig. 5a). The conditions for interference are obtained by adding two polarizers, P_1 and P_2, at 45° and 135° to the Wollaston prism, respectively; the vector diagram corresponding to a beam that crosses the prism at its center is given in Fig. 5b. The actual arrangement is shown in Fig. 5c, where the use of the same Wollaston prism as a compensator can be observed (to permit the use of an extended source). Finally, the introduction of a compensator of a known angle permits the accurate evaluation of the difference in height between two points of the observed surface (sensitivity: ~100 Å).

Another application of the same optics is the observation of the surface relief by *interference contrast*, where a very small phase shift is introduced and only the zeroth order fringe is observed. Hence, a uniform field will be observed except at those points where brighter or darker lines correspond to discontinuities or sharp gradients in the surface.

The general advantage of the light microscope investigations (compared with those of electron microscopy) is that they can be used *in situ*, i.e., the change of crystal form

---

* Classical light-optical measurement, e.g., normal microscopy and reflectance measurements, do not allow sufficiently sensitive indications of surface irregularities to be useful in future work.

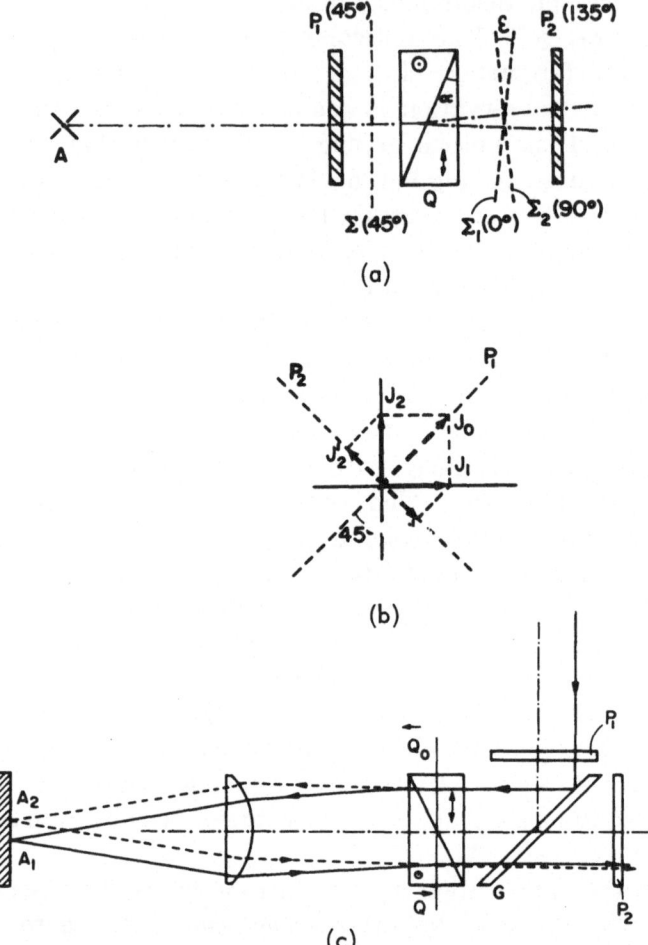

(a)

(b)

(c)

Fig. 5. (a) Schematic diagram showing working principle of Nomarski polarized interferometer; A is the source, P_1 the polarizer, Q the Wollaston prism, and P_2 the analyzer; $\Sigma(\alpha)$ are the wave fronts corresponding to the different positions in the optical system (plane polarized light at an angle α with respect to the Wollaston). (b) Vector diagram corresponding to a beam passing through the center of the prism. J_0: after passing through P_1; J_1, J_2: after passing through Q, where a splitting into two beams takes place. The projections of J_1 and J_2 on the direction corresponding to P_2 are also given, showing, in this case, destructive interference. (c) Actual system (Nomarski and Weill[75]).

with time *during the deposition process* can be observed.[76]

(ii) Electron Microscopy

This technique is usually applied indirectly, e.g., only after making replicas. Crystal growth and rearrangement occurs after the net current is switched off and, therefore, in order to obtain information about the deposit, *freezing* of the situation during deposition (e.g., via the sudden exchange of the solution by a replicamaking fluid) should be attempted. Reflection electron diffraction can also be used for the examination of metal surfaces,[87] but the *in situ* technique seems impractical due to adsorption of the electron beam by the solution. Transmission electron microscopy is also possible if a very thin substrate mounted on a support which is dissolved prior to the examination is used (the information obtained in this case concerns the bulk of the thin deposit rather than its surface).

(iii) X-Ray Diffraction

The possibility of using X-ray diffraction for the examination of a growing crystal seems worthy of consideration. That only the surface atoms are observed could be ensured by using a very low angle of incidence with respect to the surface.[77] An attempt at X-ray examination of an electrolytically growing surface has been made recently[82] and a careful analysis of the experimental possibilities seems desirable.

(iv) Ellipsometry

This technique (by which measurements are made of the changes in the polarization state of a beam of polarized light when it is reflected at a surface) has recently been developed in respect to electrochemistry.[78] It allows determination of the properties (thickness, refractive index, conductance) of films with a thickness as low as 1 Å. In metal deposition studies it would yield secondary information, e.g., concerning surface films associated with the deposition and perhaps information about inhibitor layers (it is particularly of application in film-forming reactions).[79]

3. SURFACE AND SOLUTION PREPARATION

(i) General Technique

This has been recently described.[80] The principal point is the creation of solutions which have a very high degree of purity—indeed, much higher than that used for other electrode reactions. A high vacuum system could form the basis of this technique. It is desirable to consider new experimental procedures to obtain high purity water; its possible preparation *in situ* from Pd-filtered H_2 and Ag-filtered O_2 should be explored. Gaseous HCl forms the basis of conductance in the electrolyte and the metal is introduced anodically. The solution is in contact with its own vapor only. The problem of which kind of material to use as the vessel is not yet solved. Possibly, highly purified and aged Teflon would give less total dissolution of capillary-active impurities than most glasses.[17]* Pure quartz might be the best possibility.

(ii) Controlled Substrate by Means of Electrodissolution

See Chapter 7.

(iii) Controlled Substrate by Means of Guillotine Technique

Hagyard· and Williams[81] have used a sharp knife, traveling at 700 cm/sec, to prepare a fresh surface. The method is advantageous for metals such as aluminum, whose bare surface is difficult to study because in solution it is normally covered with an oxide film.

(iv) Ion and Neutron Bombardment

Maintenance of a single crystal above the solution in

---

* However, the porosity of Teflon and, consequently, its permeability to gases constitutes an undesirable property.

vacuum compartments under ion[83] or neutron bombardment before bringing it into contact with the solution is a possibility. However, this technique does not seem to offer significant advantages over (ii), as long as the solution used in the electrodissolution technique can be exchanged for one which has not been in contact with the impure surface.

(v) Detection of Small Amounts of Organic Substances Intentionally Introduced into the Metal Surface

The present radio-tracer methods[84,85] allow the detection of surface coverages as low as 0.01. Lesser amounts can be estimated from a knowledge of the standard free energy of adsorption at higher coverages, extrapolating this to zero coverages, and use of this value in a simple isotherm to calculate the low coverage situation.

Chapter 4

Basic Models of the Atomic Movements Which Lead to Deposition

1. A DEFINITION

In order to facilitate the systematic study of the electro-crystallization process it is convenient to make a distinction between "crystal growth"—the cooperative process by which the aggregate of depositing particles build up a new crystal—and that other aspect of the same overall process which may be called "deposition," i.e., the sequence of events that *each* ion follows from the moment it arrives in the double layer until it becomes incorporated into the crystal lattice.

2. MECHANISM OF INTERFACIAL CHARGE TRANSFER

In much of the earlier work in electrode kinetics, it was thought that the atomic mechanism corresponding to the process $M^+ + e \rightarrow M$ was that of an electron jump from the metal to the positive ion, and then it seemed reasonable to refer to the process as "neutralization." However, it was soon found[89] that this name did not correspond very well to the event which occurred at the interface, particularly for metal deposition. Thus, the center of the positive ion is several angstroms out into the solution when the process occurs. Yet, inside the metal (at least for those metals, e.g., silver, in which

27

the electron gas model is regarded as that most applicable for the bulk structure) it is still an ion, and only an extra electron—not directly associated with it—is "in the metal." Thus, viewing the metal as a whole, it neutralizes the charge of the deposited ion. Because this is not "neutralization" in the simple sense of, e.g., a gaseous ion which is neutralized by an electron, it is best to speak of the process $M^+ + e \rightarrow M_{(adsorbed\ on\ surface)}$ as "charge transfer." The $M_{(adsorbed\ on\ surface)}$ could also be written as $(M^+ + e)$. However, orbital interactions between localized electrons bind the particles together in some metals so that, as $(M^+ + e)$ would not be a generally appropriate way of representing a "discharged ion," it is better to use the symbol M, with the understanding that M perhaps does not represent an atom of M.*

* M's nature is discussed further, below.

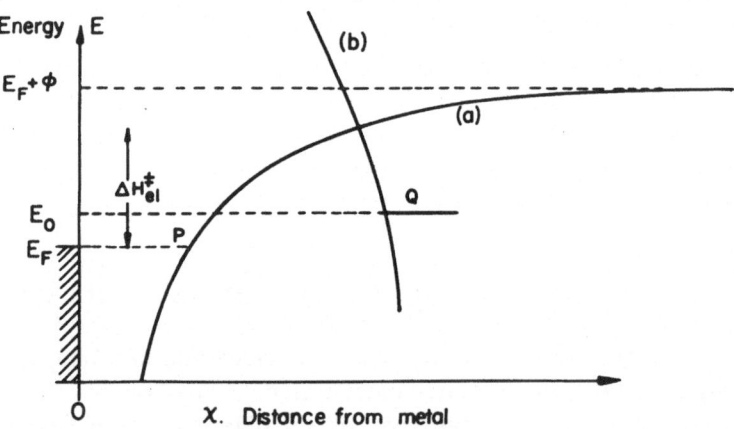

Fig. 6.   Potential energy diagram for *electron transfer* from the metal to an accepting ion in the double layer. Curve (a): Potential energy for removal of an electron from the metal; Curve (b): Potential energy for removal of an electron from the particle (atom) to form the ion. The diagram corresponds to *one* position of the particle in the diagram of Fig. 7, the points P and Q on each curve being the initial and final states for electron transfer in this figure (Matthews[94]).

It is easy to show[94] that there is a very small rate of classical electron emission *over* the energy barrier for electrons at an electrode. There must, therefore, be tunnel transfers of the electrons from metal to solution. A fundamental condition for radiationless transfer of an electron through the energy barrier (see Fig. 6) is that the energy E_F of the electron in the metal (state P on curve (a)) is equal to that E_0 in the receptor state at the other side of the barrier (state Q on curve (b)). Detailed calculation[94] shows that it is the frequency of occurrence of such acceptor states which determines the rate of transfer of the electron through the barrier to the solvated ion in solution.

The occurrence of the quantum states in the ion-solvent entity in the double layer can be considered in terms of Fig. 7, where curve A shows the variation of the energy of an ion in solution (plus an electron in the metal) as a function of the varying distance between the ion center and the

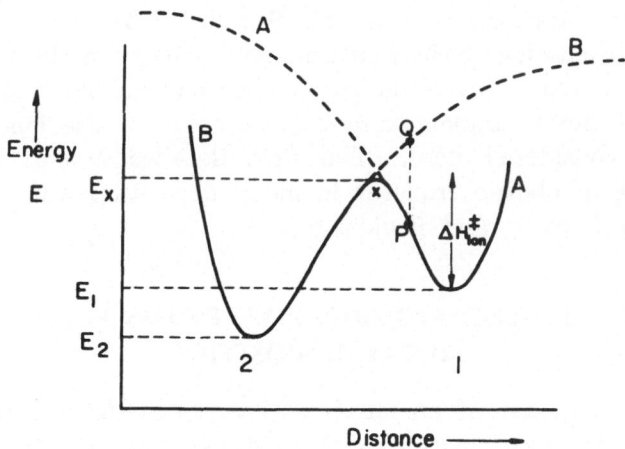

Fig. 7. Potential energy diagram for *ion transfer* at the metal solution interface. Curve A: potential energy for hydrated ion; curve B: potential energy for an adsorbed particle on the surface, after charge transfer (adatom or adion) (Matthews[94]).

effective center of a typical water molecule in its hydration sheath. Curve B shows the corresponding energy of a neutralized surface atom (plus water molecules in the double layer) as a function of the distance between the surface atom concerned and the mean position of the average surface atoms. Each configuration (or distance) corresponds to two points, P and Q, on the curves A and B, respectively. These points are the initial and final states for ·the *electron transfer* process whose energetics is depicted in Fig. 6.

Analysis of the condition for electron tunneling (Gurney,[88] Butler,[89] Bockris and Matthews[94]) shows that this is attained when the paths of the two energy–distance curves intersect (X in Fig. 7, $E_F = E_0$ in Fig. 6). The main energetic parameters which determine the probability of arrival of a particle in the solvation sheath to a configuration sufficiently stretched so that it satisfies the conditions for electron tunneling are the ionization energy of the atom concerned, the repulsion energy between the atom on the surface and the water molecules in the double layer, the heat of sublimation of the metal, and the solvation energy of the gaseous metal ion.

Other authors, particularly Marcus,[90,91] have concentrated upon a discussion of the rearrangement energy in the solvation sheath in redox reactions (*cf.* Sacher and Laidler,[92] Mott and Watts-Tobin[54])—analogous to the stretching of the ion-solvent bonds—considered here. The first detailed calculations of the rate of charge transfer in metal deposition was made in 1958 by Conway and Bockris.[52]

3. THE ALTERNATIVE PATHS FOR METAL DEPOSITION

The transfer of an ion from its hydration sheath[95,110] in solution to some position on the metal surface (see Figs. 1 and 8) can occur in several different ways.

1. Transfer of the ion from its hydration sheath in solution to a position on a surface plane of the metal, where the ion is still partly hydrated (see below) and would have a

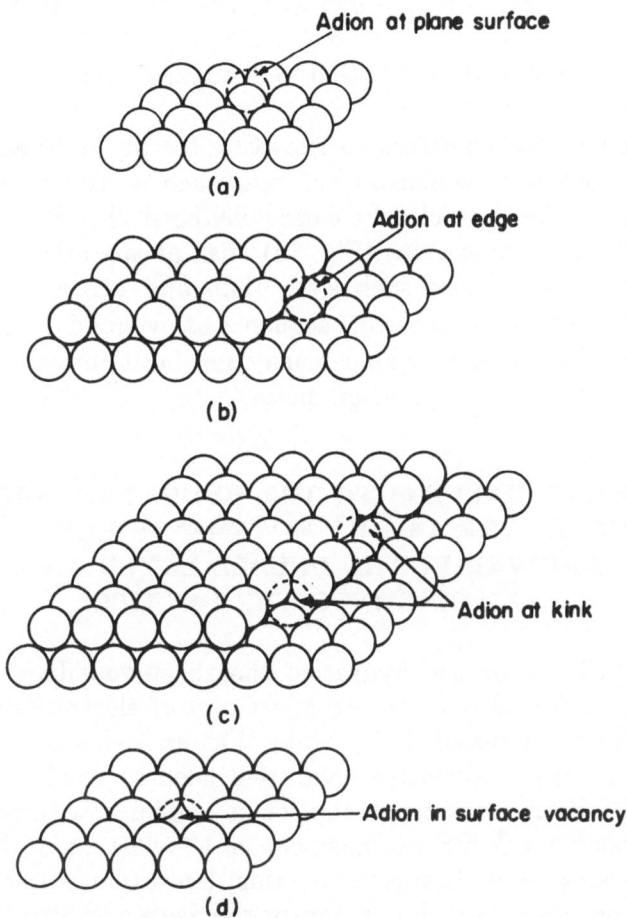

Fig. 8.   Diagrammatic representation of arriving ions (dotted circles) at
(a) planar surface (C.N. = 3); (b) edge (C.N. = 5); (c) kink site
(C.N. = 6); (d) vacancy (C.N. = 9) on a {111} surface of a crystal
(hydrating molecules not shown) (Conway and Bockris[53]).

coordinate number (C.N.) of 3* with respect to the metal
atoms (Fig. 8a).

2.   Transfer from the hydration sheath in solution to a
site at the edge of a step (C.N.=5, Fig. 8b).

* The numerical values of the C.N.'s mentioned in this section correspond to those for a {111} plane in a face center cubic metal.

3.   Transfer from the hydration sheath to a kink (C.N. = 6, Fig. 8c).

4.   Transfer from solution to a surface vacancy (C.N. = 9, Fig. 8d).

Such direct transfers to growth sites are not necessarily the only ways in which an ion can reach a site of growth. A priori, there is clearly more likelihood that the ion will first land at a plane site (Fig. 1a) than at any other, because there are many more such sites than any other. Figure 1 shows a picture of a possible sequence of events during deposition and illustrates how an ion may get built into the crystal and finally form a "piece of metal."

4.   WHAT HAPPENS TO THE HYDRATION SHEATH DURING THE ATOMIC MOVEMENTS BETWEEN ARRIVAL IN THE DOUBLE LAYER AND BUILDING INTO THE LATTICE?

Knowledge of the hydration sheath surrounding a metal ion in solution is not the strongest part of electrochemistry.* However, it is reasonably certain that an oriented layer of water molecules (primary hydration sheath) resides with an ion in solution, as long as the ion is sufficiently small (less than about 1.7 Å for a monovalent ion).[97] The major difficulty in assessing what happens to this hydration water during deposition is in lack of knowledge of the degree of directionality of the ion-water bond. Because of this, it is assumed here that the bond is purely coulombic, i.e., nondirectional, which is certainly true for the hydration of metal ions of Group IA and IIA,[98] and perhaps for IB.

Thus, whether hydration water remains attached to a particle after the charge transfer stage depends upon whether some ionic character is retained by the particle. If this is so, the type of model which would be expected is shown in

* The model both for water on the electrode and that surrounding an ion in solution is discussed by Andersen and Bockris.[96]

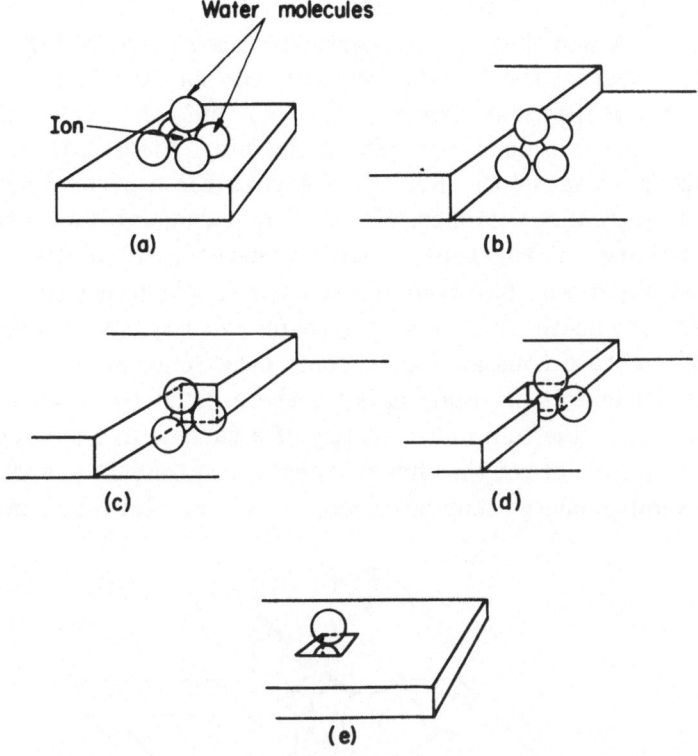

Fig. 9. Diagrammatic representation of possible modes of hydration of transferred adions at various sites upon the metal surface (Conway and Bockris[52]).

Fig. 9. The number of hydrating molecules retained decreases as the depositing particle replaces water molecules in its co-ordinating sphere on the surface by other metal atoms.

5. IS CHARGE TRANSFER TO ALL TYPES OF SITES ON THE METAL SURFACE EQUALLY PROBABLE?

It is of considerable importance to know the most likely type of site on the substrate at which charge transfer takes place, and if the rate constant (or the heat of activation) for the transfer to each type of site is known, the path of the reaction can be predicted. Calculations on this point were

first published by Conway and Bockris[52] in 1958. The basic
situation which has to be considered is shown in Figs. 10
and 11. During the transfer process, the ion has to *displace
part of its hydration sheath*. The amount of displacement
which has to be done depends on the site. Clearly, the more
the displacement of the hydration sheath for a given type of
site, the greater the energy needed to surmount the barrier
for discharge at the type of site concerned. This does not
necessarily mean that transfer to the site which most distorts
the water sheath need have a greater energy of activation
(or lower rate constant) than some other one (at which the
distortion energy is less), because there could be a compensation, e.g., due to a lower energy of a particle in a particular
type of *final* state of the charge transfer process ($M^+ + e \rightarrow M_{ads}$).
For example, energy may be released by the greater coordination

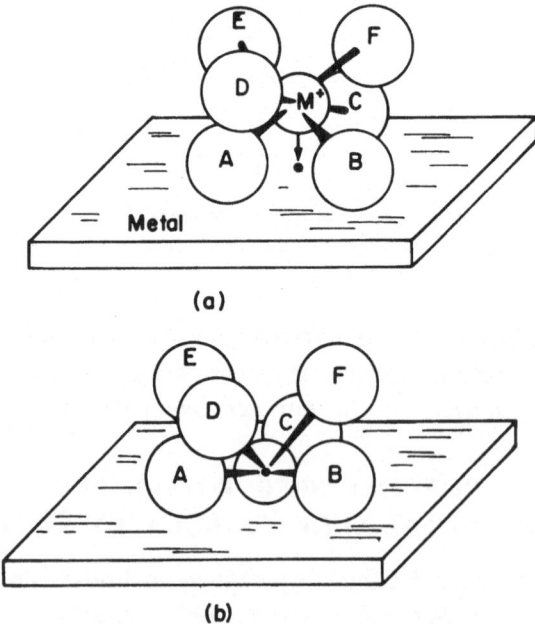

(a)

(b)

Fig. 10. Model of symmetrical transfer of a divalent ion to the metal
surface: (a) initial state; (b) after displacement (outer primary
hydration layer is not shown) (Conway and Bockris[53]).

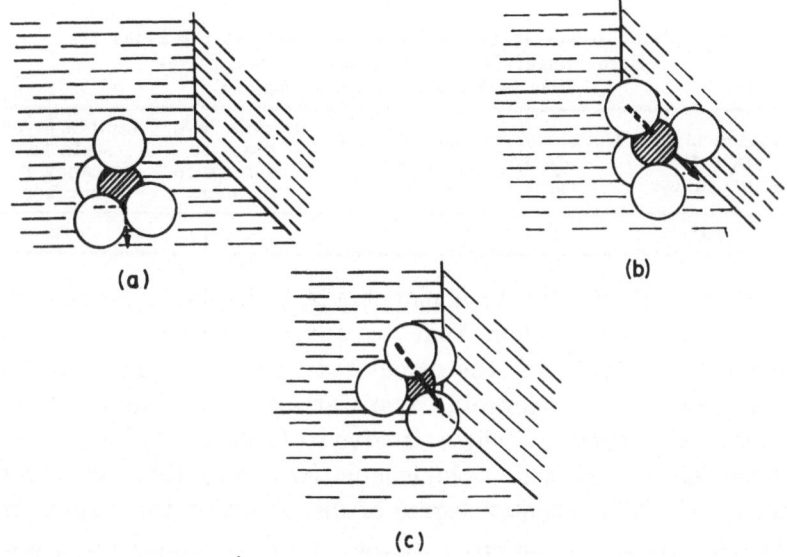

Fig. 11. Situation of hydrated monovalent ions in the double-layer state at (a) a plane surface site, (b) an edge, and (c) a kink site. Arrows indicate direction of ionic transfer (Conway and Bockris[53]).

of the transferred ion with the substrate's metal atoms in a site (e.g., a kink site) at which the distortion is particularly large, and thus compensate the effect of the large distortion energy.

Conway and Bockris' calculations showed that a tendency to such compensation existed but was insufficient, to compensate the distortion energy and that *the heat of activation for transfer to a planar surface site was much less than that to an edge (of a growing layer), to a kink, or to a vacancy in an edge or in a surface.** Moreover, the difference in calculated heats of activation is great, so that the rate

* The result substantially effects discussions of the path in metal deposition. Mott and Watts-Tobin[54] assumed that the rate constants for transfer were the same to all types of sites. Fleischmann and Thirsk,[42] neglecting the influence of the dehydration energy, suggested that charge transfer was more likely at a kink than at a planar site, because of the lowering of the energy of the final state by greater coordination.

Table 2

**Calculated Heats of Activation in kcal/mole at the Point of
Zero Charge for Direct Transfer from Silver
Ions to Various Types of Adions**

Double layer state to planar surface	Double layer state to edge	Double layer state to corner (kink)	Double layer state to vacancy in an edge	Double layer state to vacancy in a surface
10	21	35	> 35	≫ 35

constant for transfer to a planar site is likely to be at least 10^6 times greater than that to any other type of site. A comparative table of the calculated activation energies for a univalent ion undergoing transfer across the double layer to different types of sites is given (Table 2). Illustrations of the models on which the calculations were based are given in Fig. 11. The greater degree of distortion of the hydration sheath (and hence increase in energy of the transition state) in edge and kink landings is shown (see Fig. 9 and compare the coordination numbers of the adsorbed metal particles given in the annotation of Fig. 8).

6.   CAN "TWO-ELECTRON TRANSFERS" TAKE PLACE?

In considering metal deposition reactions of the type $Cu^{2+} + 2e \rightarrow Cu$, a question arises as to whether "two-electron charge transfer" can occur, i.e., whether two charges may be transferred *in one step* (which might be a rate-determining step). The calculation[53] indicates that this is not possible, because the heat of activation (>100 kcal/mole) is so high that the rate constant for this type of transfer (i.e., charge transfer by the model described in Section 2 of this chapter, but involving two electrons ejected from the metal and reaching the transition state simultaneously) is negligible. Hence, $Cu^{2+} + 2e \rightarrow Cu$ must take place in separate steps.*

---

* In fact, it was shown[61] that the rate-determining step at potentials not too near the reversible potential was the one electron charge transfer, $Cu^{2+} + e \rightarrow Cu^{+}$. The result was confirmed by Nekrassov *et al.*[99] Hurlen[100,101] contradicted it, but in order to explain his experimental
(*Continued on next page*)

7. ADATOM OR ADION?

An *a priori* calculation of the degree of ionic character which the bond between metal and the surface makes after charge transfer is difficult. But it is possible to make calculations on the extreme assumption that the particles on landing from the solution are adsorbed atoms (i.e., zero residual charge), and then to compare the value of the calculated heat of activation with that obtained from calculations in which the adsorbed particles are assumed to retain a dipolar character with respect to an electron in the metal (i.e., the particle on the surface is an "adion," rather than an "adatom").

The results of such calculations[52,53] show that the heat of activation to form silver adatoms from aqueous solutions is about 30 kcal/mole, and thus is it far higher than the corresponding heat of activation (about 10 kcal/mole) calculated for the formation of silver adions, i.e., for the different model in which a degree of charge on the ion is maintained after "charge transfer" (the adsorbed ion and the concomitant electron form a dipole on the surface). In the latter case, some hydration water remains attached to the adsorbed particle, and the stabilization of the "final state" in the transfer reaction due to this residual hydration (Fig. 9) is the physical basis for the lower heat of activation in charge transfer to *adions rather than to adatoms.*

A similar conclusion was reached independently, and on quite different grounds, by Gerischer.[41]

8. PATH AND RATE-DETERMINING STEP IN DEPOSITION: THEORETICAL INDICATIONS

If, as calculations at present seem to show, it is *improbable* that transfer of an ion from solution to the metal surface

results (which are not conclusive and disagree with those of other workers) he must invoke a chain mechanism involving "active sites" as the chain carriers, the physical nature of which is not clear, and that involves deep problems in their interpretation.

occurs to a kink site, if the heat of activation for transfer
to a planar site is lower than that to any other on the metal
surface,* and if the product of the charge-transfer reaction
is an adsorbed particle which retains some degree of charge,
the principal question which further theoretical calculations
must answer is: Which of the consecutive steps—charge
transfer, surface diffusion, transfer from surface site to edge
of growth site, transfer from site on edge to kink site—is
associated with the greatest difficulty, i.e., is the rate-
determining step in the deposition process?

The potential energies of partly hydrated adions in each
of the above situations can be calculated, and the stretching
constants which indicate the increase in energy of a particle
when it moves out of an equilibrium position can be taken
into account. What is found for the few calculated situations[53]
(Ag^+, Cu^{2+}, Ni^{2+}) is that the charge transfer step is rate-
determining when the potential is relatively negative, but
that near the reversible potential the uncertainty in the
numerical values is too much to distinguish between rate-
determining charge transfer and rate-determining surface
diffusion. The two following steps, associated with transfers
at growth sites, seem to be relatively fast, or "reversible.†"

Such theoretical approaches are rough and orientive.
They start as several guesses of equal status and end up as
"backed-up guesses," and "backed-down guesses." The conclusions are tentative estimates because of the crudeness of
the models and the poor definition of the quantities involved.
But they are helpful in designing experiments which confirm
or deny the suggestions to which the calculations give rise.

---

* The coverage of the metal surface with water dipoles,[54,102,103] the
orientation of which depends on the potential at the metal-solution interface, must not be forgotten in detailed considerations of the landing of
ions on a surface and their diffusion across the surface to growth sites.

† However, it has been found that for Cu electrodes in chloride
solutions incorporation at growth sites is apparently rate-determining,
at least at very low overpotentials.[181]

Chapter 5

Transients

1. ORIGIN OF TRANSIENT MEASUREMENTS
IN METAL DEPOSITION STUDIES

The origin of this resides in the single peculiarity which the study of metal deposition offers to electrode kinetics, namely, that the substrate changes with time during the reaction. Reference has been made to Volmer's suggestion.[24]

2. THE MEHL AND BOCKRIS ANALYSIS

Rojter, Juza, and Polujan[25] were the first to record transients in metal-solution exchange reactions. However, they did not attempt to analyze their results in terms of a detailed mechanism. Mehl and Bockris[13,38] observed that the transient in the deposition and dissolution of Ag from $AgClO_4$ in $HClO_4$ exhibited a much longer rise time than that calculated if the charge transfer reaction were assumed to be rate-determining and double layer charging the only process responsible for the transient behavior. A longer delay in attaining a steady rate of deposition would occur if the number of adions on the surface during the transient did not adjust rapidly to a new steady-state condition as the ions arrived (in a cathodic transient). Hence, either surface diffusion or aggregation to a growth site must be, for the condition of the transient to which reference is made, relatively slow compared with the charge transfer reaction. The differential equation governing surface diffusion is

$$\frac{\partial \bar{c}}{\partial t} = \frac{i_F}{zF} - \bar{v} \tag{1}$$

where $\bar{c}$ is the mean adion concentration on the surface, $\bar{v}$ is the mean rate of removal of adions from the unit surface area (by surface diffusion), and i_F is the net Faradaic current density. The functional dependence of $\bar{v}$ on $\bar{c}$ was assumed by Mehl and Bockris in a first approximation to be linear, i.e.,

$$\bar{v} = v_0 \frac{\bar{c} - c_0}{c_0} \tag{2}$$

which also implies that the adion concentration at a step is the equilibrium one (fast incorporation at kinks). By introducing (2) in (1) and integrating with the initial condition $\bar{c}(0) = c_0$, the time dependence of the mean adion concentration when a current step is applied to the electrode is obtained:

$$\frac{\bar{c}(t) - c_0}{c_0} = \frac{i_F}{zFv_0}\left[1 - \exp\left(-\frac{v_0}{c_0}t\right)\right] \tag{3}$$

The time dependence of the overpotential was obtained from (3) using the $i - \eta$ equation for charge transfer corrected to take into account the variable concentration of adions, i.e.,

$$i_F = i_0\left\{\exp\left[-\alpha_c \frac{zF}{RT}\eta(t)\right] - \frac{\bar{c}(t)}{c_0}\exp\left[\frac{\alpha_a zF}{RT}\eta(t)\right]\right\}$$

The explicit expression for $\eta(t)$ was obtained for $|\eta| < RT/\alpha zF$ by expanding the exponentials and neglecting second-order terms:

$$i_F = i_0\left\{1 - \frac{\alpha_c zF}{RT}\eta(t) - \frac{\bar{c}(t)}{c_0}\left[1 + \frac{\alpha_a zF}{RT}\eta(t)\right]\right\}$$

$$= i_0\left\{1 - \frac{\bar{c}(t)}{c_0} - \left[\alpha_c + \frac{\bar{c}(t)}{c_0}\alpha_a\right]\frac{zF}{RT}\eta(t)\right\}$$

Assuming that $[\alpha_c + \alpha_a\bar{c}(t)/c_0] \approx 1$, introducing (3) for $[\bar{c}(t) - c_0]/c_0$, and solving for $\eta(t)$ yields

$$\eta(t) = -\left\{\frac{RT}{zF}\frac{i_F}{i_0} + \frac{RT}{z^2F^2}\frac{i_F}{v_0}\left[1 - \exp\left(-\frac{v_0}{c_0}t\right)\right]\right\} \quad (4)$$

Finally, i_F can be equated to the total current density passing across the electrode if the double layer charging current is neglected, i.e., if $t \gg \tau_{DL} = RTC_{DL}/zFi_0$. Equation (4) can then be written as

$$\eta(t) = \eta(\infty) + \frac{RTi_F}{z^2F^2v_0}\exp\left(-\frac{v_0}{c_0}t\right) \quad (5)$$

with

$$\eta(\infty) = -\left(\frac{RT}{zF}\frac{i_F}{i_0} + \frac{RT}{z^2F^2}\frac{i_F}{v_0}\right) \quad (6)$$

Therefore, the (linear) plot of log $[\eta(t) - \eta(\infty)]$ vs. t will give v_0/c_0 from the slope and v_0 from an extrapolation to $t = 0$, allowing also the evaluation of i_0 from (6).

Values of c_0 for Ag electrodes due to various authors are shown in Table 3.

The Mehl and Bockris analysis[13] grew not only from Volmer's suggestion, but also from papers by Lorenz,[40,104] in which surface diffusion as a rate-controlling step in metal deposition was suggested, and that of Gerischer and Tischer,[105]

Table 3
The Surface Adion Concentration on Silver Electrodes at the Reversible Potential as a Function of Surface Preparation

Surface preparation	Author	c_0 (mole/cm^2)*
Quenched in H_2 atmosphere	Mehl and Bockris[13]	90×10^{-11}
Scrape in solution	Gerischer[41]	15×10^{-11}
Undefined	Lorenz[70]	7×10^{-11}
Quenched in He	Despic and Bockris[62]	3×10^{-11}
Prepared *in situ* by anodic pulse	Despic and Bockris[62]	160×10^{-11}
Electrodeposited from melt	Reddy[164]	90×10^{-11}†

* A close-packed layer of atoms has a surface density of 230×10^{-11} mole/cm^2.

† Ag electrode in molten AgCl–LiCl–KCl.

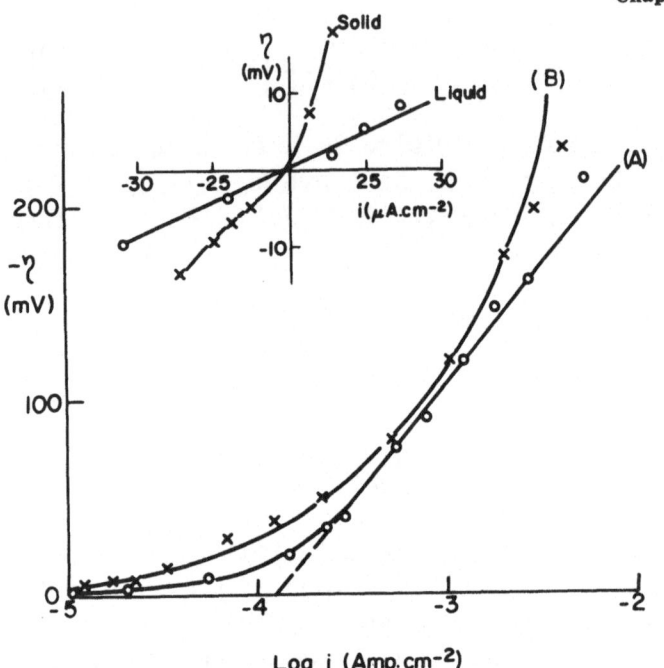

Fig. 12.   Experimental points and theoretical η-log i curves for Ga
electrodes.   (A) theoretical curve for pure charge transfer
control; (B) theoretical curve for a mechanism involving surface
diffusion; open circles: experimental points for liquid Ga
electrode (30°C); crosses: experimental points for solid Ga
electrode (28°C) (Bockris and Enyo[65]).

in which the concept of mobile atoms on the metal surface was
formulated in relation to the mechanism of electrodeposition
of silver, and their participation in the building up of a
concentration overpotential was suggested.

A strength of the Mehl and Bockris analysis was that
it was found possible to abstract i_0 (the rate of the charge
transfer reaction at the reversible potential) from the analysis
of the potential–time curve and make a comparison with i_0
obtained otherwise (agreement ~30%).   However, it is
desirable to get a *direct confirmation* of the rate-determining
character of a step occurring after the charge transfer and
on the way to lattice building.   A good way to do this is to
measure the kinetics just on either side of the melting point

of the metallic electrode; liquids do not build up crystals with emergent dislocations and growth sites and, hence, the comparison of the behaviors immediately exposes the effect of steps connected with surface diffusion on transfer from a planar site to a growth site. The experiment was done by Bockris and Enyo with Ga electrodes,[65] and it showed clearly the greater difficulty of deposition on the solid when the temperature difference between liquid and solid electrodes was negligible (Fig. 12).

3. BETTER APPROXIMATIONS IN THE ANALYSIS OF TRANSIENTS

The Mehl and Bockris analysis involved rudimentary assumptions and approximations. The growth sites were assumed to be in equilibrium with the planar surface in the immediate vicinity, and the concentration gradient for adion diffusion from any point between the growth sites was assumed to be constant. The algebra was carried out with assumptions concerning values of $\eta (<10\,\mathrm{mV})$ and c_{ad} ($\approx c_0 \ll$ number of ad-sites on surface) which restrict the range of applicability.

Fleischmann and Thirsk[42] attempted a more detailed analysis of the non-steady-state conditions. They obtained a complete solution for the potentiostatic case,* but as in the earlier and more rudimentary treatment, it was still subject to the limitation of not being applicable to crystal *growth*, i.e., it considered a *fixed* distance between growing steps. The potential dependence of the distance between steps and its influence on the long-time, steady-state, i–η curve was examined for both the surface diffusion and the direct-transfer models. They also attempted to obtain a more accurate solution of the galvanostatic transient than the one obtained by Mehl and Bockris, but it led to an integral equation

---

* This case, unlike the galvanostatic one, admits an exact treatment.

which could not be solved explicitly for $\eta(t)$; a linear approximation of the transient was obtained which was very near to that of Mehl and Bockris.* A third kind of transient measurement is also discussed by Fleischmann and Thirsk, namely, the determination of the AC impedance of a growing surface and its frequency variation. They also suggest the possibility of using an "AC potentiostatic transient" method, i.e., analyzing the non-steady-state portion of the i–t response of the electrode to a potential function of the form

$$\eta = 0 \qquad \text{for} \quad t \leqq 0$$
$$\eta = \eta_0 \sin \omega t \qquad \text{for} \quad t > 0$$

A third approximation in the analysis of metal deposition processes by relaxation methods is that of Damjanovic and Bockris,[106] who also studied the influence of the dislocation concentration upon the kinetic law. It was shown that the transition from surface diffusion control to charge transfer control with increasing overpotential (a feature of the experimental results[61,67]) followed expectedly because of the potential dependence of the local anodic partial current, which is the process competing with surface diffusion in the removal of adions from a given point on the planar surface. More precisely, let N be the density of dislocations intercepting the surface, D the diffusion coefficient of adions on the surface, and $p = (i_0/zFc_0) \exp[(1 - \beta)zF\eta/RT]$. It was found that when the frequency factor (ND) for arrival of adions to a kink site by surface diffusion is larger than the frequency factor (p) for escape of adions to the double layer by anodic charge transfer, then charge transfer is rate-controlling; whereas, when ND/p is less than about 0.2, surface diffusion is rate-controlling. The ratio ND/p increases with increasing absolute value of the overpotential ($\eta < 0$ for cathodic processes) and, hence, a transition from surface diffusion control to charge transfer control is observed.

* This was corrected to be exactly that of Mehl and Bockris in a later publication.[69]

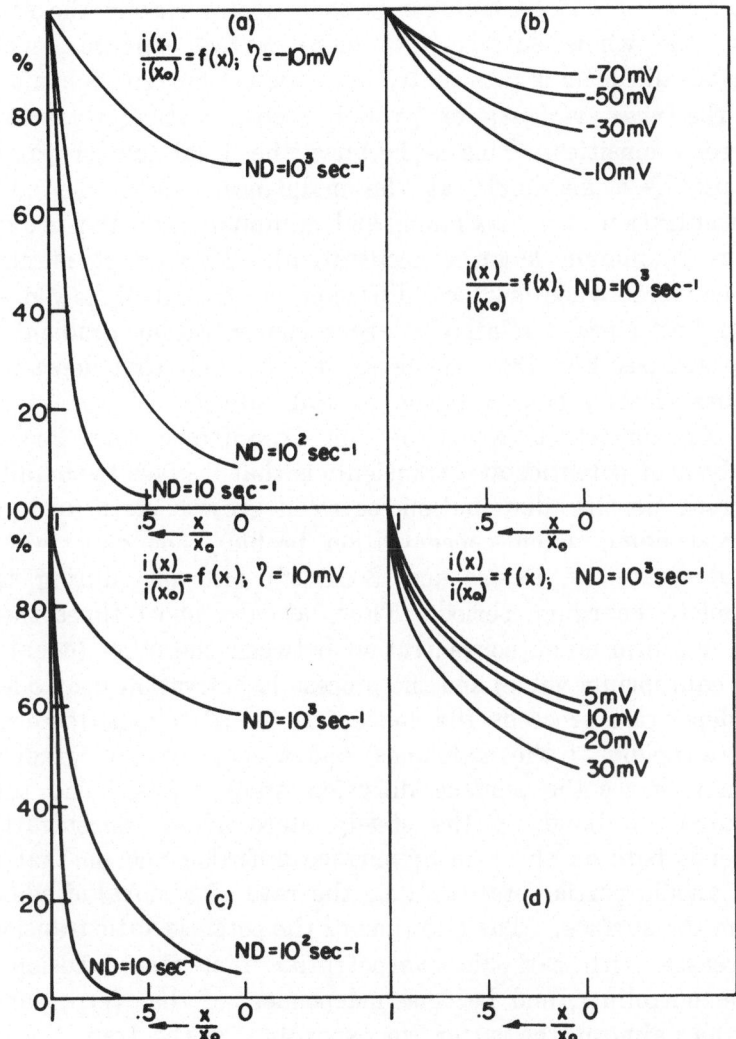

Fig. 13. Variation of steady-state local net current density with distance from growth step ($x/x_0 = 0$ at midplane; $x/x_0 = 1$ at growth step) for various η and ND values; (a) and (b) cathodic, (c) and (d) anodic (Damjanovic and Bockris[106]).

On surfaces with a high density of dislocations, diffusion of adions is relatively unimportant to rate control, since ions reaching the surface will diffuse quickly to the steps. The net rate of charge transfer is then approximately the same over the whole surface. At more perfect surfaces, when surface diffusion is rate controlling, the net current is confined to the *near vicinity* of growth steps, particularly at low current densities. This is because the local anodic partial current is a maximum at the midplanes, where the adion concentration is a maximum, and minimum near the growth steps (minimum adion concentration). This effect is more pronounced when surface diffusion is in control (low over-potential) since a relatively larger concentration gradient is required (see Fig. 13). However, it is unlikely that deposition occurs directly from solution to kink sites.[53]

An important aspect of the Damjanovic and Bockris analysis of potentiostatic transients is that it gives the detailed current density distribution between growth steps (and the corresponding adion concentration profile), not only in the steady state (*cf.* Despic and Bockris[62]) but also during the complete charging period. Thus, at very short times there is a uniform adion concentration between the steps (equal to the equilibrium value) and the process is activation controlled; as deposition proceeds the concentration of adions increases (more rapidly at the midplanes) and a concentration gradient (which drives the surface diffusion step) arises. The time required to build up the steady state adion concentration depends both on the rate of surface diffusion and on that of the anodic partial process, i.e., the rate of removal of adions from the surface. The rise time of the potentiostatic transient increases with cathodic overpotential if surface diffusion is rate-controlling, and becomes independent of the overpotential at the value of the latter corresponding to the transition to charge transfer control. The rise time at higher overpotentials (i.e., under charge-transfer control) depends, in a first approxi-mation, only on the density of dislocations,* and increases as the rate constant for surface diffusion decreases.

Knowledge of the detailed time and position dependence

* But some overpotential dependence can be expected as a result of the change of "activity" of the dislocations with overpotential.[42,107]

of the current density between steps may be of importance in a calculation of the frequency of nucleations, or perhaps in a quantitative analysis of the effect of adsorbed species on the deposition process.

This work[106] also gave rise to theoretical expressions for the dependence of nucleation velocity upon potential and the defect concentration, in agreement with work earlier performed by Vermilyea.[58]

The long time transient behavior of a metallic electrode subjected to a current pulse was analyzed qualitatively by Vermilyea.[45] The analysis was based on the spiral growth model: the screw dislocations are initially relaxed, i.e., straight steps originate from them; when current starts passing through the electrode (and crystal growth occurs), those steps acquire certain curvature, and growth spirals are formed. The distance between steps then decreases until a steady state is reached with respect to the morphology of the growing spirals. If it is assumed that there is partial surface diffusion control of the rate of deposition, it follows that there is an initially large overpotential—corresponding to a relatively large distance between steps; a decay in the overpotential would then be observed over a period of a few seconds—corresponding to the shortening of the distance between steps as the growth spirals wind up.

Budewski *et al.*[180] utilized Ag single crystal microelectrodes to study the long-term electrocrystallization process. They obtained long-time potentiostatic transients that were interpreted assuming that the initial substrate consisted of a dislocation-free atomically smooth surface on which two-dimensional nucleation occurs, followed by lateral growth until completion of a layer. A simple physical model (random nucleation, lateral growth current proportional to edge length) resulted in good agreement with the experimentally observed transients.

Chapter 6

Discussion of Some Basic Questions Concerning Path and Rate-Determining Step

1. THE CHARGE-TRANSFER STEP

The general nature of this has been outlined,[92,94] and the essence is that of tunneling of electrons to an activated state produced by stretching of the ion-solvent bonds. The majority of the heat of activation is contributed by the latter.[94] The often raised question as to why the symmetry factor has the value 1/2 can be answered as follows: It is not exactly 0.5 but the factors which determine β do make it in practice between about 0.4 and 0.6.* Thus, in the usual interpretation of the symmetry factor,[62] a variation in the applied potential produces a relative shift of the potential energy curves with respect to each other; this shift results in a change in the free energies of activation for the forward and backward reactions which is related to the slopes of the potential energy curves at the point of intersection. More precisely,

$$\beta = \frac{m_i}{m_i + m_f}$$

where m_i and m_f are the slopes of the potential energy curves of initial and final state, respectively, at the point of inter-

---

* Indeed, theories of β which yield $\beta = 0.50$ are suspect, for the experimental facts give "near to 0.5."

section. In most cases (reactions with relatively high heats of activation), the point of intersection of the two potential energy curves lies in the linear region of the Morse curves; in this case the slopes m_i and m_f are proportional to the Morse constants for ion–solvent and adion–metal surface interactions, respectively. Substitution of the range of known values for spectroscopically determined Morse constants of bonds suggests that $0.4 < \beta < 0.6$. Only when the point of intersection of the two curves corresponds to the nonlinear region of one of them is the symmetry coefficient very different from 0.5. In fact, it can then hypothetically approach zero— the end of electrodic control of the rate of charge transfer.[62]

Some authors state that "dehydration" of cations is rate-determining, but this view is not tenable because the heat of activation for total dehydration before transfer of an electron to the transition state is about one-half the first ionization energy of the ion—considerably greater than that calculated by the detailed process discussed by Conway and Bockris.[53] Hence, the rate constant for a discharge process involving the formation of a dehydrated ion is negligible.*

The ions which take part in the charge-transfer process are seldom simple, because simple cations of Groups IA and IIA in the Periodic Table occur well above H in the electrochemical series so that H_2 tends to be evolved first if attempts are made to deposit these metals; most other cations are complex forming. For example, in the deposition of Ga from solution, the following is the probable path of the charge transfer reaction:

$$H_2GaO_3^- + e \xrightarrow{\text{slow}} HGaO_2^- + OH^-$$

$$2HGaO_2^- + e + H_2O \rightleftharpoons Ga + H_2GaO_3^- + 2OH^-$$

---

* The origin of such a suggestion lies in an attempt to explain extremely small values of β obtained by the method of Faradaic rectification.[108] A more likely explanation for these is the general explanation for small β values put forward in 1960 by Despic and Bockris.[62] Their model indicates that $\beta \ll 1/2$ will be observed if the reaction is sufficiently fast. That is, indeed, the case for reactions examined by the method of Faradaic rectification.

2. THE TWO PATHS: VIA THE SURFACE OR DIRECTLY TO KINK SITES

As indicated above, a point of importance concerns the alternative paths:

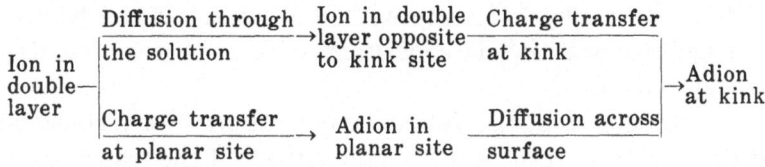

Therefore, the essential question becomes: Is there solid evidence for surface diffusion as a rate-determining step for *any* one system? If there is, it seems likely that the surface diffusion, rather than the solution diffusion, path is general, even though, in many circumstances (see below), surface diffusion is not *rate-determining*. The essential evidence is the following:

1. Galvanostatic rise times in $Ag^{13,41}$ and Cu^{61} deposition are much longer than would be expected if charge-transfer control were rate-determining, *but this is only so at low overpotentials*.

2. Values of the adion concentration at the reversible potential (calculated from experiment on the assumption of a surface-diffusion model) agree in order of magnitude when determined by three different methods (two different DC transient and one AC approach[70]). The values are of a reasonable order of magnitude ($\theta \approx 0.1$).

3. It is possible to calculate i_0 from equations for η as a function of time during a DC galvanostatic transient, assuming surface-diffusion control. The value obtained is in satisfactory agreement with i_0 determined from the steady-state situation at higher overpotentials when the charge transfer reaction is rate-controlling and the evaluation of i_0 is unambiguous.

4. The dependence of the adsorption capacitance on the concentration of ions in solution yields a partial charge on the adion[41] which is consistent with the model theoretically expected for an adparticle undergoing surface diffusion.

5. There is a change over—from surface-diffusion control to rate control by charge transfer—upon an increase of over-potential above approximately 50–100 mV (at least for certain methods of surface preparation).[62] This is the theoretical expectation if the path does involve surface diffusion.

6. The change of transients and steady-state behavior upon melting for Ga[65] is consistent with a degree of surface-diffusion control.

7. Independent theoretical investigations[40,53] are consistent *only* with charge transfer at *planar* sites. If the transfer were to occur at kink sites, there would be a linear relation between current density and overpotential, which is not observed.

Therefore, evidence for the functioning of a path involving surface diffusion is good.

Fleischmann and Thirsk[42,69] imply that the path in electro-deposition would be more likely in the few examined cases to involve direct transfer to growth sites preceded by diffusion in the solution. However, their deduction is based upon the premise that there is a *higher* (instead of lower) rate constant for direct discharge to a growth site. Apparently, this suggestion arose from qualitative considerations of the greater coordination number of the adion at a growth site (i.e., the implied lowering of the energy of the final state in the charge transfer reaction due to greater bonding). Therefoer, it appears that Fleischmann and Thirsk thereby neglected the change of energy along the reaction coordinate in the ion's approach to the activated state caused by the greater distortion necessary during transfer to growth sites than at the planar sites (see Fig. 11). The ratio of probability of discharge at a growth site to that at a planar site is, hence, likely to be very much less than that assumed by Fleischmann and Thirsk; the tentative conclusions which they make upon such an as-sumption are, consequently, not soundly based.*

* The estimation of this ratio depends on the relative values of the energy of solvation and sublimation. The calculations suggest that a heat of sublimation much above 80 kcal/mole would be necessary to cause charge transfer to kink sites to become faster than that to planar sites.

Vermilyea criticized the suggestion of surface-diffusion control of Lorenz,[40] Mehl and Bockris,[38] and Gerischer[41] on the grounds that the electrodes used in their investigations should be highly rough, therefore obviating the necessity for a surface diffusion stage.* He attributes the long rise time observed to the effect of impurities which would cause non-uniform growth and, consequently, local concentration polarization. However, this hypothesis seems unlikely in view of the large number of successful correlations between the predictions of surface diffusion theory and observed behavior (*cf.* above). It is also difficult to explain how a presumably small amount of adsorbed impurities would hinder the direct deposition onto the very large number of sites connected with an atomically rough surface.[55]

3. THE GENERALITY OF RATE CONTROL BY THE CHARGE-TRANSFER STEP AT HIGH CURRENT DENSITIES

Evidence has been given which makes it appear likely that ions transfer to surface planes and diffuse along the surface to the growth sites; that if the overpotential and the density of growth sites are sufficiently small, surface diffusion is in rate control; and, finally, that rate-controlling surface diffusion turns into rate-controlling charge transfer at higher overpotentials. Why is the last trend a general one?

An initial naive approach would be to expect that if a diffusion step were at any time rate-controlling, increase of rate would tend to make it increasingly controlling. However, two factors tend to counteract this expectation:

1. Not all the emergent dislocations on the surface of the metal give rise to *active* growth steps. The fraction which is active increases with increase of overpotential (see Chapter 7).[107] But, as the fraction of growth sites which is active increases, the effective length through which an adion

* This contention would be strictly correct only in the case of atomically rough surfaces.

has to move decreases, i.e., the rate of surface diffusion for a given degree of supersaturation increases and, hence, the tendency to rate control by surface diffusion decreases.

2.   There is a general reason why a change from surface-diffusion to charge-transfer control should occur with increasing cathodic potential even if the activity of the growth sites were

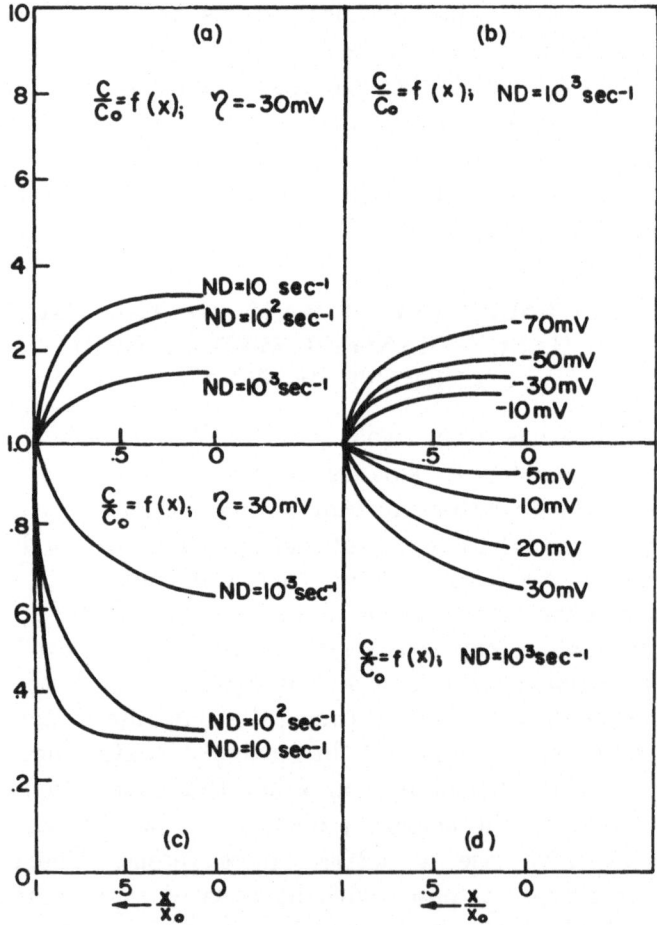

Fig. 14.   Variation of steady-state adion concentration with distance from growth step ($x/x_0 = 0$ at midplane; $x/x_0 = 1$ at growth step) for various η and ND values; (a) and (b) cathodic, (c) and (d) anodic (Damjanovic and Bockris[106]).

independent of potential*: as the overpotential becomes more cathodic, the rate of the (local) anodic process decreases; the time that a particle spends on the surface after charge transfer and the probability of that particle reaching a step and being incorporated at a growth site increases. Since this process occurs by surface diffusion, the increase in probability is equivalent to an increase in the rate constant for surface diffusion which, therefore, will eventually cease to be rate controlling.

4. UNDER SURFACE-DIFFUSION CONTROL, DEPOSITION DOES TAKE PLACE PREFERENTIALLY NEAR A GROWTH SITE

It has been argued in this chapter that the general mechanism of metal deposition involves surface diffusion (Chapter 4), and that deposition directly to growth sites is relatively improbable. However, it must be noted[62] that the current (in the sense of net rate of arrival of new material from solution) is not uniform for the cathodic case under surface diffusion control, but is small in the midplanes and larger near the growth sites (because c_{ad} is large in the midplanes). Hence, for surface diffusion control, preferential charge transfer near to, but not at, growth sites is likely. As the charge transfer becomes rate-controlling at higher overpotentials, this effect would relax and charge transfer would tend to take place almost uniformly at all points between growth steps (see Figs. 13b and 14b).

---

* Reference is made to cathodic deposition. A rather different situation with respect to this point obtains in the dissolution.[106]

Chapter 7

Deposition and Dissolution Kinetics as a Function of the Initial Substrate

The study of the variation of the velocities of exchange of metal atoms with the solution and that of their diffusion to growth sites for metal deposition is analogous to that of catalyses in other types of reactions. The stage of development is early; however, several interesting matters already emerge.

1. VARIATION OF THE ACTIVITY OF A SURFACE WITH TIME

It has been known for many years[33] that there is a variation in the activity of a freshly formed surface with time. More recently, observations of the rate of deposition have been made as a function of the life time of the surface.[68] The surface is prepared by anodic dissolution, and the fresh surface is allowed a life of 10^{-2}–10^3 sec before the rate of deposition on it is measured by means of a transient technique. This study shows a striking variation of the activity of the surface with its age (Fig. 4).

The surface retains an initial high activity for about 1 sec, and then a rapid decrease of activity occurs during the next 10^2–10^3 sec.

Several models can be considered for this type of change of activity with time.*

The following models are considered:

1. During anodic dissolution, spirals unwind, i being larger at points with a smaller radius of curvature. Pits may develop. On cessation of the net dissolution current, dissolution or deposition occurs locally at peaks or pits, respectively, so that the surface tends to become smoother, i.e., its real area is less with the increase of time. The rate constant per apparent square centimeter would then tend to decrease with time. However, such effects would be reflected in the capacitance measurements, and these show values independent of time after cessation of dissolution.

2. After cessation of the anodic current, impurities from the solution (e.g., any dissolved from the electrode surface) may diffuse back to the surface and adsorb thereon, thus blocking growth steps.

Calculation on the basis of this model[68] shows that

$$\eta_t = \eta_{t=0}\left[\left(1 - \frac{N'_{t=\infty}}{N}\right) + \frac{N'_{t=\infty}}{N}\exp\frac{D't}{K^2}\operatorname{erfc}\frac{(D't)^{1/2}}{K}\right]^{-1}$$

where $\eta_{t=0} = -RTi/8(zF)^2Dc_0N$, N'_t is the number of growth steps poisoned at time t after creation of a fresh surface, K is the coefficient of adsorption of the impurity,[112] D' is the diffusion coefficient for the impurity in the solution, N is the total number of growth steps, i is the constant current density applied during the galvanostatic transient, and surface diffusion control has been assumed for the deposition process, D and c_0 being the diffusion coefficient and equilibrium concentration of adions, respectively. This model (Curve a, Fig. 15) represents well the observed results in all except one important aspect: it requires the adion concentration at the equilibrium potential c_0 to be constant with time. It is observed that a

---

* There is an analogous change for most electrode reactions, not only for metal deposition. It has been made the basis of a continuous activation procedure in electrochemical energy converters.[111]

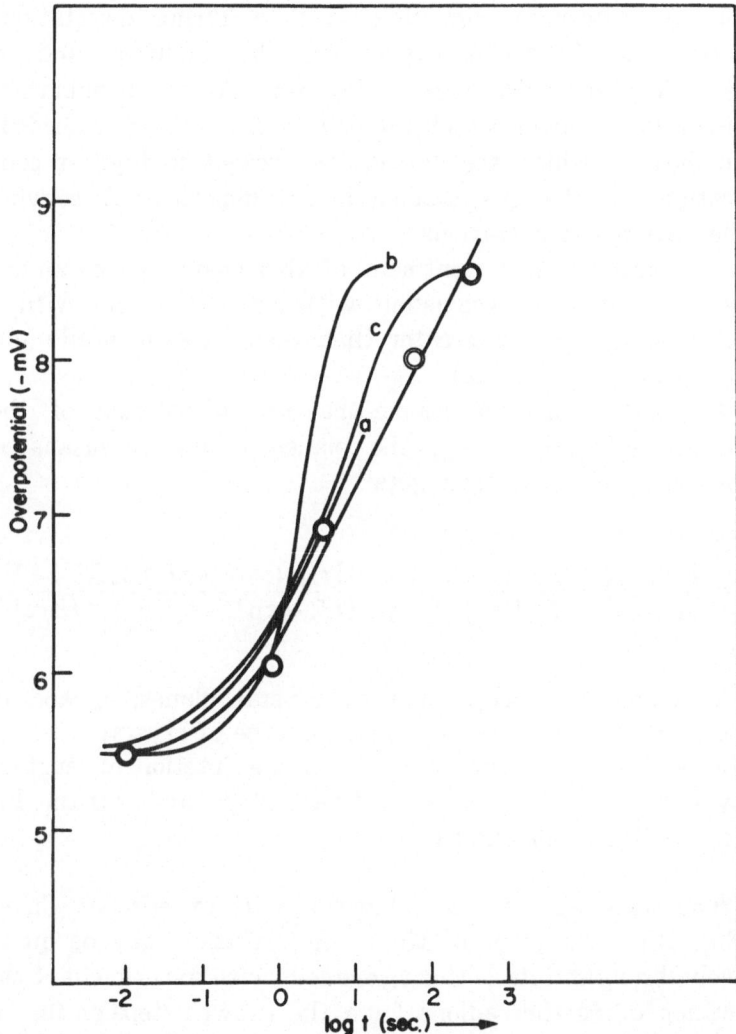

Fig. 15. Steady-state overpotential for a galvanostatic transient started after a time t of contact between fresh surface and solution. Experimental points for Cu electrode. Curves (a), (b), and (c): Calculated for three different models for the change of activity of a fresh surface with time (Bockris and Kita[68]).

sharp fall occurs in the adion concentration during the first 1 sec of exposure of the surface to the solution, and this is not consistent with the model.

3. An alternate possibility is that during dissolution, crystal faces of various indices may be produced, and on these, adions adsorb to various degrees. At the termination of dissolution, during which the new crystal faces are formed, the planes on which the adions are present in highest concentration may dissolve, and metal ions deposit on those with smaller adion concentrations.

The mathematical treatment of this model[68] gives an expression qualitatively consistent with experiment but with a slope η vs. $\log t$ much greater than that experimentally observed (Curve b, Fig. 15).

4. Lastly, one can assume that the adjustment of the adion concentration among the planes occurs by means of adion diffusion. One then gets

$$-\eta_t = \frac{RT}{zF}\ln\left\{1+\frac{\{1+[(c_{A,t=0}/c_{B,0})-1]f_A\}\cdot\{\exp(-zF\eta_{t=0}/RT)-1\}}{1+[(c_{A,t=0}/c_{B,0})-1]f_A\cdot\exp[(-2f_A/x_0)\sqrt{Dt/\pi}]}\right\}$$

where A and B are two kinds of crystal planes, c_A and c_B are the adion concentrations on each type of crystal face, $c_{B,0}$ is the equilibrium value of c_B, f_A is the fraction of surface of type A, x_0 is the distance between steps, and D is the diffusion coefficient of adions.

The model agrees reasonably well with experiment (Curve c, Fig. 15). The slope of the η–$\log t$ relation may be qualitatively brought into better agreement with experiment if the formation of further adions from the growth steps on face A is taken into account, the effect being to slow down the equilibration of adion concentration among the crystal faces. It is not unplausible that the rearrangement occurs by surface diffusion, because the i_0 value on electrode surfaces in which c_0 is high, is small compared with the i_0 values on He-prepared electrodes, i.e., adions on the more populated surfaces go into solution less readily than those on the less populated surfaces.

2. THE EFFECT OF THE POTENTIAL DEPENDENCE OF THE ACTIVITY OF THE EMERGENT DISLOCATIONS ON THE KINETICS OF DEPOSITION[107]

Consider a system in which transient examinations of the kinetics of metal deposition are being made, and suppose that the rate-controlling reaction is that of surface diffusion. A basic assumption of this model is that there is equilibrium between atoms incorporated at the growth sites and adions on the surface immediate to growth sites. Physically, this means that when an ion diffuses and meets a growth step (e.g., one originating from the rotatory growth movement of a spiral), the latter always readily *accepts* the new adion and continues to grow.

In reality, a certain number of the growth sites which originate from emergent dislocations will react in this way, and a number will not. It has been shown[44,106] that there is a critical length r_c such that a growing layer must have a radius of curvature r larger than r_c in order to continue growing. This phenomenon is the two-dimensional equivalent of the increase in vapor pressure of liquid droplets as their radius of curvature decreases.

The radius of the critical nucleus is given by

$$r_c = \frac{A\gamma_{\text{edge}}}{\rho_s RT \ln\left[c(\eta)/c_0\right]} \tag{7}$$

where A is the atomic weight of the metal, γ_{edge} is the edge free energy (two-dimensional equivalent of the surface tension), ρ_s is the surface density in a two-dimensional nucleus, $c(\eta)$ is the maximum possible value of the surface adion concentration at the overpotential η, and c_0 is the adion concentration at the reversible potential (i.e., in equilibrium with a straight step). The value of $c(\eta)$ will be related to the overpotential by

$$\frac{RT}{zF} \ln \frac{c(\eta)}{c_0} = -\eta \tag{8}$$

because the situation involves an *equilibrium* between ions in solution at an overpotential η and adions on the surface at a concentration $c(\eta)$. It follows from (7) and (8) that as $|\eta|$ increases, the value of the critical radius at which the growing layer accepts adions for growth is decreased.

If it is assumed that the steps present on the surface arise mainly from pairs of emergent screw dislocations, the average step length will be given by

$$\bar{l} = \left(\frac{1}{N_{dis}}\right)^{1/2} \tag{9}$$

where N_{dis} is the density of dislocations intercepting the surface. However, not all the steps arising from pairs of dislocations will be active; given a pair of dislocations at a distance l apart, the radius of curvature of the step joining them decreases as the step grows, and will pass through a minimum equal to $l/2$. Therefore, only those steps for which $l \geqslant 2r_c$ will be able to accept adions for continuous growth. Assuming a random distribution of dislocations, the fraction of steps on the surface with length greater than $2r_c$ will be[107,178]

$$P = \exp\left[-\left(\frac{2r_c}{\bar{l}}\right)^2\right] \tag{10}$$

introducing (7), (8), and (9) in (10),

$$P = \exp\left[-\frac{4N_{dis}A^2\gamma_{edge}^2}{z^2F^2\rho_s^2}\frac{1}{\eta^2}\right] \tag{11}$$

This equation indicates that the fraction of steps which are *active* increases very rapidly as the overpotential increases. The net effect is to make the rate constant for surface diffusion higher as the overpotential increases—one reason why there should be a tendency to change to rate-controlling charge transfer with increasing overpotential.

3. THE EFFECT OF CHANGE OF CONCENTRATION OF EMERGENT DISLOCATIONS IN THE INITIAL SUBSTRATE

The effect of potential upon the activity of the growth sites at low overpotential is relatively small. It is of interest to consider how the kinetics of metal deposition depends upon grosser changes in the initial substrate, such as those which are introduced by cooling the metal at different rates from the liquid state (and, consequently, varying the dislocation density). A theoretical study of this has been carried out recently by Damjanovic and Bockris.[106] The general expression for the steady-state current is

$$i = i_0 \left\{ \exp\left[-\frac{\alpha z F \eta}{RT} \right] \right.$$
$$\left. - \exp\left[(1 - \alpha)\frac{z F \eta}{RT} \right] \right\} \sqrt{\frac{4ND}{p}} \tanh \sqrt{\frac{p}{4ND}} \qquad (12)$$

where N is the density of dislocations, D is the diffusion coefficient of adions, and

$$p = \frac{i_0}{zFc_0} \exp\left[(1 - \alpha)\frac{zF}{RT}\eta \right]$$

For $ND \geqslant p$, $\sqrt{4ND/p} \tanh \sqrt{p/4ND} \approx 1$ and equation (12) reduces to

$$i = i_0 \left\{ \exp\left[-\frac{\alpha z F \eta}{RT} \right] - \exp\left[(1 - \alpha)\frac{z F \eta}{RT} \right] \right\}$$

This accounts for the gradual transition from rate-controlling surface diffusion at low cathodic overpotentials (p large) to rate-controlling transfer step at more cathodic overpotentials (p small), as experimentally observed by Despic and Bockris.[62] As the overpotential becomes increasingly negative, the adion concentration increases, so that $|dc/dx|_{x=\pm x_0}$ increases (see Fig. 14) and, hence, the surface diffusion rate increases to a non-rate-controlling value. The overpotential at which transition from rate-controlling surface diffusion to rate-controlling transfer during deposition occurs, depends on N and D. At a more perfect crystal surface (N is small) the transition will occur at more negative overpotentials. For surface diffusion to be the rate-controlling step, it is necessary that [$cf.$ equation (12)]

$$ND < p = \frac{i_0}{zFc_0} \exp\left[(1 - \alpha)\frac{zF\eta}{RT}\right]$$

or, at low overpotentials

$$ND < \frac{i_0}{zFc_0}$$

Thus, for silver electrodes in a solution of $0.2N$ AgClO$_4$, for instance, with $i_0 = 0.1$ A/cm², and $c_0 = 10^{-10}$ mole/cm², surface

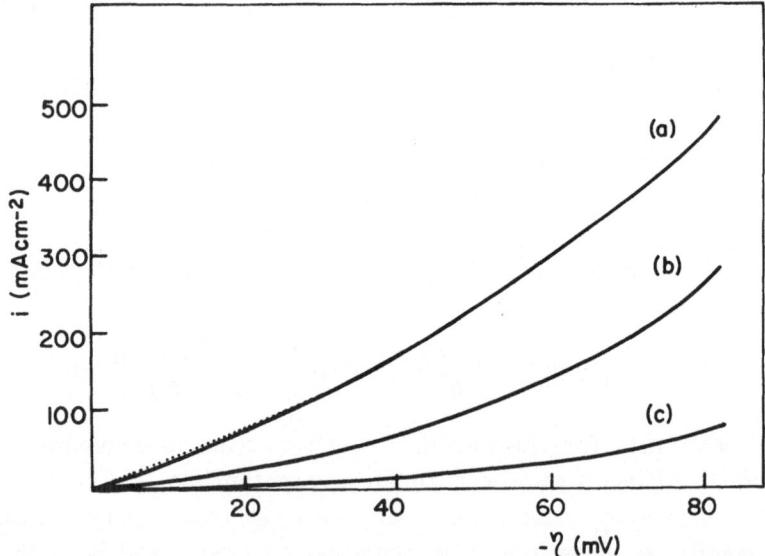

Fig. 16. Theoretical i–η curves for deposition (charge transfer–surface diffusion mechanism) for different values of the surface diffusion parameter ND: curves (a), (b), and (c) correspond to $ND = 10^4$, 10^2, and 10 sec^{-1}, respectively; the dotted curve corresponds to pure charge transfer control ($ND = \infty$). In all cases $i_0 = 0.1$ A/cm², and $c_0 = 10^{-10}$ mole/cm² (Damjanovic and Bockris[106]).

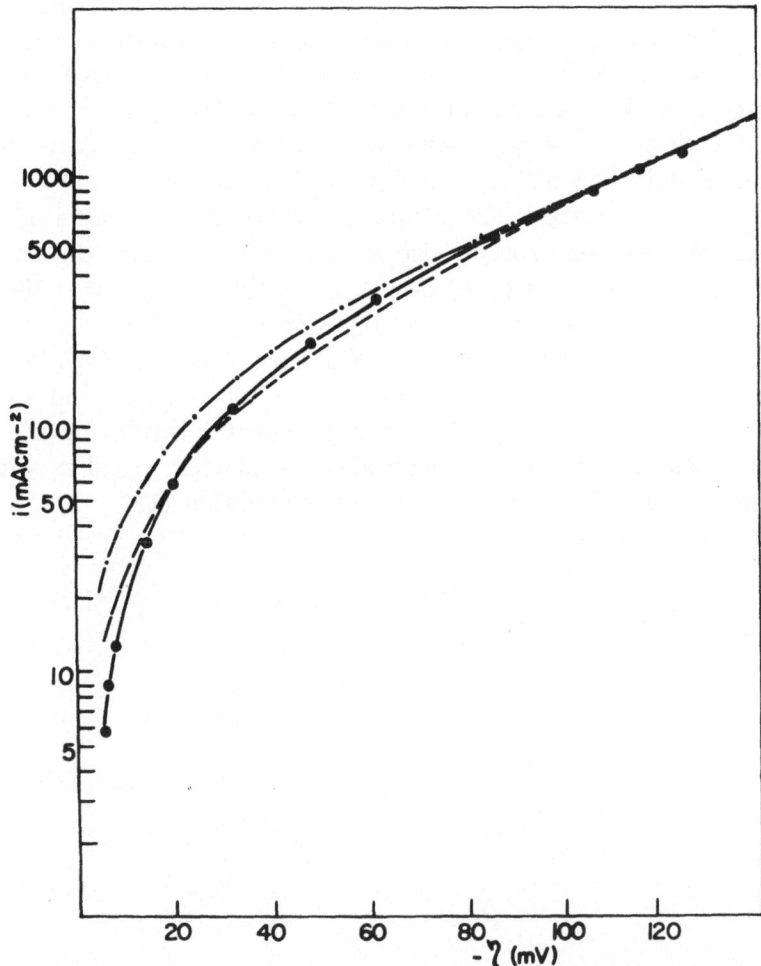

Fig. 17. (Solid line) log i-η plot of experimental results of Mehl and
Bockris for Ag deposition. (Dashed and dotted line) closest fit
for a pure charge transfer mechanism. (Dashed line) charge
transfer-surface diffusion mechanism with $i_0 = 0.12$ A/cm², $c_0 =$
10^{-10} mole/cm², $ND = 3.4 \times 10^3$ sec^{-1} (Damjanovic and Bockris[106]).

diffusion will be rate-controlling if ND is less than 10^4 sec^{-1}.
In fact, from Fig. 16 it can be seen that for $ND = 10^4$ sec^{-1}
and $\eta = -10$ mV, the steady-state current density is only a
few percent smaller than it would be if transfer were the

rate-controlling step. It is possible to estimate the value of the diffusion coefficient of silver adions in the system used by Mehl and Bockris. The best fit with their experimental data (see Fig. 17) can be obtained if in equation (12) ND is $3.4 \times 10^3 \, \text{sec}^{-1}$. A reasonable value of N for their electrodes can be taken as $10^9 \, \text{cm}^{-2}$, and then D is approximately equal to $5 \times 10^{-6} \, \text{cm}^2/\text{sec}$. The accuracy with which the value of D can be obtained depends also on the accuracy with which c_0 is determined. Here, c_0 is taken as $10^{-10} \, \text{mole/cm}^2$. With decreasing N values, surface diffusion will tend increasingly to become rate-controlling. In Fig. 16 a family of i–η curves is given for various ND values, $i_0 = 100 \, \text{mA/cm}^2$, and $c_0 = 10^{-10} \, \text{mole/cm}^2$. In Fig. 17 the experimental results of Mehl and Bockris[13] for silver electrodes are plotted together with the theoretical curve obtained using equation (12).

Chapter 8

Why Do Some Crystals Grow in Filamentary Shapes?

1. CRYSTAL GROWTH, DENDRITIC GROWTH, AND THE STUDY OF THE MECHANISM OF ELECTROCRYSTALLIZATION

It is now necessary to turn from "metal deposition" studies—the study of the mechanism by which ions move from

Fig. 18.   Electromicrograph of an Ag dendrite, grown from 0.1 M AgNO$_3$ at 100 mA/cm^2 (Matthews, Mutucumarana, and Wilman[158]).

the solution to the growth sites on the metal surface—and study the growth itself. This is a more complex task and it is convenient to begin with the single example in which the situation is fairly well understood, i.e., that of systems which grow in a dendritic manner.

Dendritic shapes (see Fig. 18) grow frequently in nature. Snow-flakes are dendritic.[113,148] When metals crystallize from the the liquid, they frequently do so by growing dendritically from the cool surface.[114] Most metal deposits can be made to grow dendritically. The choice of this particular ·type of crystal growth as the *initial* one for study is good because the growth rate and shape are easily measurable, the emergence of a dendrite from a crystal face is a striking exhibition of crystal growth, and this type of growth occurs in so many situations that its explanation must be a rather general one.

2. FACTS ABOUT DENDRITIC GROWTH ON ELECTRODES

In spite of its frequent occurrence, both on electrodes and in growth associated with zero net current, there is little quantitative information concerning the growth of dendrites.[115,116] The summary of phenomenology given here arises principally from observations[47] on the *potentiostatic* growth of Ag from molten KNO_3–$NaNO_3$ containing various amounts of $AgNO_3$. A small Ag sphere was used as the substrate, and efforts were made to maintain temperature uniformity at the relatively high temperatures used (>300°C). The effect of impurities upon the kinetics also was examined to some extent. The facts are as follows:

1. Whether dendrites grow or not depends primarily upon the ratio between current density (referred to the substrate area) and concentration of the solute. Dendrites will not grow, for a given macrocurrent density, if the concentration of Ag^+ is sufficiently high; correspondingly, for a given concentration of Ag ions, they will not grow if the current density is sufficiently low (Fig. 19a).

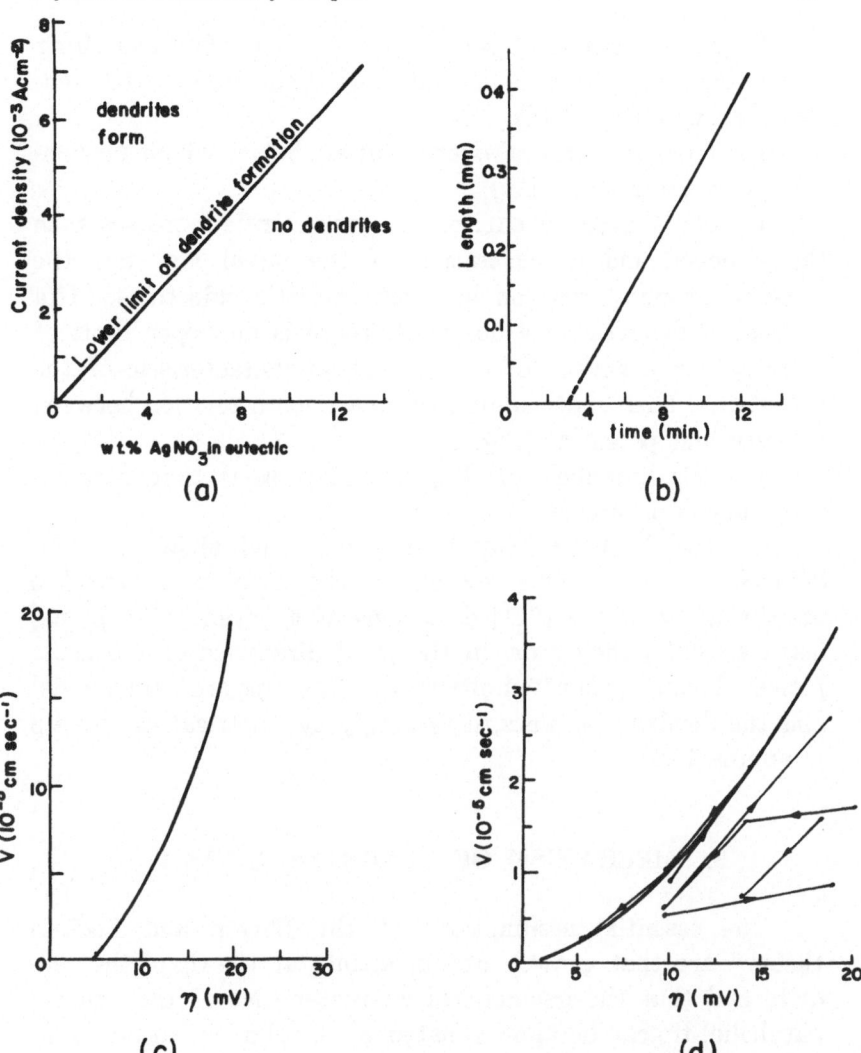

Fig. 19.   Experimental facts concerning electrolytic growth of Ag dendrites from a molten system: (a) critical current density for dendritic growth as a function of concentration of $AgNO_3$ in the melt; (b) growth of dendrite, showing induction time ($\eta = 11\,mV$, $c_{AgNO_3} = 4.5\%$); (c) maximum velocity of dendritic growth as a function of potential, $c_{AgNO_3} = 1.08 \times 10^{-4}\,mole/cm^3$; (d) velocity of dendritic growth as a function of potential, showing velocity for individual dendrites (points connected by straight lines) and maximum velocity (curve), $c_{AgNO_3} = 0.35 \times 10^{-4}\,mole/cm^3$ (Barton and Bockris[47]).

2.   If conditions in 1. are favorable, then, after switching on the current, there is an initiation time before the dendritic growth begins (Fig. 19b).

3.   There is a critical overpotential, below which no dendrite will grow (Fig. 19c).

4.   For a given dendrite, the growth rate increases with the potential and concentration of the metal ion, but the velocity–potential relation has characteristic inflections. If a number of dendrites are observed, there is an upper limit of velocity for a given potential which is characteristic of the potential. This upper limit lies on a smooth relation between velocity and potential (Fig. 19d).

5.   Side branches (*cf.* Fig. 18) grow some time *after* a new surface is created.

6.   The dendritic growth is planar and shows a well-defined crystallographic structure: the stem and branches usually grow in the [211] directions of a given {111} plane; less frequently they grow in the [110] directions of the same plane. Twinning is often observed. The tips are paraboloidal and the dendrite becomes increasingly symmetrical as the tip is approached.

3.   MECHANISM OF DENDRITIC GROWTH

The essential assumptions of the Barton and Bockris theory[47] are that growth occurs mainly at the tip of the dendrite and that the geometrical characteristics of the (nearly parabolic) tip can be approximated by a spherical model (Fig. 20a). This last assumption allows the formulation of the simple mathematical relations between the radius of curvature at the tip r, the observed *local* current density i,* and the different kinds of overpotentials. These relations, can be summarized by the following equations:

* For any surface this is given by $i = (zF/V)v$, where V is the molar volume and v is the (local) linear rate of growth of the surface in a direction normal to the substrate.

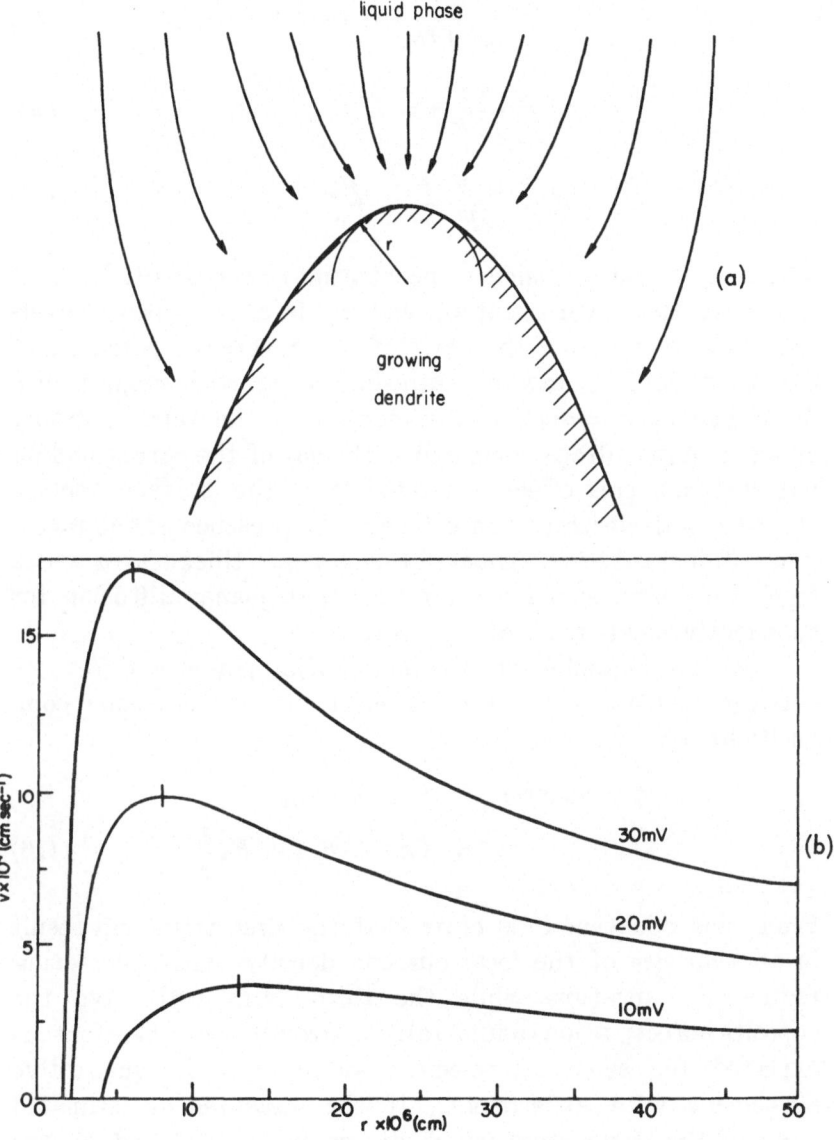

Fig. 20. (a): Parabolic and spherical models of a dendrite tip. (b): Theoretical curves showing relation between tip radius and velocity for dendritic growth at various overpotentials ($T = 581°K$; $D_{Ag^+} = 1.4 \times 10^{-5}$ cm²/sec; $c_{Ag^+} = 5 \times 10^{-4}$ mole/cm³; $i_0 = 50$ A/cm²; $\gamma = 2 \times 10^3$ ergs/cm²) (Barton and Bockris[47]).

$$\eta_d = \frac{RT}{(zF)^2 Dc_b} \, ir = K_d ir \qquad (13)$$

$$\eta_a = \frac{RT}{i_0 zF} \, i = K_a i \qquad (14)$$

$$\eta_k = \frac{2\gamma V}{zF} \frac{1}{r} = \frac{K_k}{r} \qquad (15)$$

where η_d is the diffusion (concentration) overpotential, η_a is the activation overpotential, and η_k is a "curvature" overpotential arising from the shift of the reversible potential of a curved surface toward cathodic values with respect to a flat electrode ($r = \infty$) (*cf.* the increase in the vapor pressure of small liquid drops compared with that of the corresponding flat surface); this effect is related to γ, the surface tension of the metal-electrolyte interface. The presence of the factor r in (13) instead of the usual diffusion-layer thickness δ arises from the onset of spherical rather than planar diffusion for sufficiently small radii of curvature.

The total (applied) overpotential (i.e., potential difference between test and flat reference electrode of the same composition) is

$$\eta = \text{constant} = \eta_d + \eta_a + \eta_k$$
$$= K_d ri + K_a i + K_k \frac{1}{r} \qquad (16)$$

From this equation it is clear that the first term will result in an *increase* of the local current density with decreasing radius of curvature while the third term will have the opposite effect; a maximum rate of growth can therefore be expected for some intermediate value of r in qualitative agreement with experiment. If the measured or estimated value of the parameters which determine K_d, K_a, and K_k are introduced in (16), a rate of growth *vs.* radius relation of the general shape shown in Fig. 20b is obtained, which clearly indicates the impossibility of a rate of growth higher than a given value ($v_{\max}$). From the theoretical relation (16) it is

Table 4

Comparison of Theory and Experiment for Velocity of Dendritic
Growth as a Function of Overpotential and Concentration

Differential coefficient	Concentration (mole/cm^3)	Theory* (cm/sec V)	Experiment (cm/sec V)
$\left(\dfrac{\partial v_{\max}}{\partial \eta}\right)_{\eta=3\text{mV}}$	3.5×10^{-5}	1.9×10^{-3}	1.0×10^{-3}
	10.8×10^{-5}	6×10^{-3}	3×10^{-3}
	76×10^{-5}	4×10^{-2}	7×10^{-2}
Differential coefficient	Concentration (mole/cm^3)	Theory (cm^4/sec mole)	Experiment (cm^4/sec mole)
$\left(\dfrac{\partial v_{\max}}{\partial c_b}\right)_{\eta=10\text{mV}}$	——	1.1	0.5

* Assuming $\gamma = 2000 \text{ erg/cm}^2$, $i_0 \rightarrow \infty$.

possible to obtain the dependence of $v_{\max}$ on η, K_d, K_a, and
K_k, i.e., a functional relation of the type

$$v_{\max} = F_{\gamma, i_0, D}(c_b, \eta)$$

where γ, i_0, and D are parameters which must be calculated
or estimated from existing data. In Table 4 a comparison
of the theoretically predicted and experimental values is given;
it is assumed that the activation overpotential is negligible
$(i_0 \rightarrow \infty)$.

From Fig. 20b it is also clear that the radius for maximum
rate of growth increases sharply with decreasing overpotential
(in fact, at low η the theory predicts a relation of the type
$r_{\text{opt}} \propto 1/\eta$). Consequently, a rapid fall in the current density
corresponding to the dendritic growth process would occur at
some low value of η. But at a given electrode, there are
usually several parallel electrode reactions which compete, and
only the most rapid is observed. With the sudden fall of the
dendritic growth rate at low overpotentials, indicated by the
present theory, the current would probably be sustained by
an alternative process, e.g., pyramid growth on the substrate
crystal.

The initiation, and the associated latency period, also
follow from the present hypothesis. The primary reason for

preferential growth at a tip is the enhanced diffusion conditions thereto, caused by the onset of spherical rather than linear diffusion. This effect cannot become effective while the growing surface remains in a region in which the concentration gradient is controlled by the diffusion layer of the larger electrode: the tip must form its own diffusion layer *outside* the overall diffusion layer. In accordance with this model for initiation, microscopic observations of the crystal surface at low potentials showed[47] that the period between the switching on of a constant potential and the beginning of dendritic growth was accompanied by faceting of the surface followed by a building of small prismatic outgrowths (*cf.* Aten and Boerlage[28]). These entities have to penetrate the diffusion layer characteristic of the spherical substrate before dendritic growth begins.

Semiquantitative interpretations of the maintenance of the shape of the tip radius, and the side arm growth, also can be made in the model stated.

A more sophisticated treatment based on the same principles was introduced by Hamilton.[117] Its main features are: (*a*) the model for the dendrite surface is a paraboloid of revolution rather than a spherical surface; (*b*) allowance is made for the fact that the boundary conditions apply on a *moving surface* (i.e., the advancing dendrite tip); (*c*) it assumes as a postulate the conservation of shape of the growing dendrite. On these bases a slight improvement is achieved in the fit to the experimental data of Barton and Bockris, but an abnormally low value of the surface tension must be assumed ($\gamma \approx 185\,\mathrm{ergs/cm^2}$) together with a rather high value for the exchange current density. The theory provides a fairly quantitative interpretation for the kinetic aspects of dendritic growth but not for its morphological ones.

In 1958 Price, Vermilyea, and Webb presented a theory for electrolytic whisker growth which is based on the inhibition of crystal growth by adsorbed impurities and takes into account the essentially kinetic character of the adsorption process. They assumed[50] that growth (once nucleation has occurred) proceeds through the lateral propagation of layers. Under these conditions an adsorbed species could either prevent further growth—when the mean distance between adsorbed particles ($\bar{l}$) is less than twice the radius of the critical nucleus (r_c)—or become trapped into the crystal (buried)—when $\bar{l} > 2r_c$. In the second case, the surface concentration of the adsorbed species will be given by the steady-state condition applied to the rates of arrival and trapping of impurity molecules at the surface:

$$\frac{d\Gamma_i}{dt} = \frac{D_i c_{i,b}}{r} - \frac{iV}{zF}\frac{\Gamma_i}{h} = 0 \tag{17}$$

The first term in the expression for $d\Gamma_i/dt$ gives the maxium rate at which impurity molecules reach the surface (per unit area) and become adsorbed, and the second term gives the number of moles buried per unit time and unit surface by the new crystal grown. D_i is the diffusion coefficient of the impurity in the solution; $c_{i,b}$ is the concentration of impurities in the bulk of the solution; r is the radius of curvature of the whisker tip; V is the molar volume of the metal; z is the valence of the metal ion; Γ_i is the surface concentration of impurities (in mole/cm^2); and h is the height of an adsorbed impurity molecule (in a direction normal to the surface).

The assumptions leading to (17) are:

1. The rate of adsorption is controlled by (spherical) diffusion in the solution.
2. The probability of an adsorbed particle being desorbed rather than incorporated in the crystal is negligible.

If it is assumed that the particles are uniformly adsorbed over the whole surface, the mean distance between them $\bar{l}$ is simply related to Γ_i by

$$\bar{l} = (N_0 \Gamma_i)^{-1/2} \tag{18}$$

where N_0 is Avogadro's number. From (17) and (18),

$$\bar{l} = K_1 \left(\frac{r}{c_{i,b}}\right)^{1/2} i^{1/2} \tag{19}$$

with

$$K_1 = \left(\frac{V}{zFN_0D_ih}\right)^{1/2}$$

The radius of the critical nucleus is related to the overpotential by (see Chapter 7, Section 2.)

$$r_c = \frac{A\gamma_{\text{edge}}}{\rho_s}\frac{1}{\eta} \tag{20}$$

A linear relation was assumed between current density and overpotential:

$$\eta = K_{\text{tr}}i \tag{21}$$

Price, Vermilyea, and Webb proceeded at this point to identify η in equation (20) with η in equation (21), thus obtaining a relation between r_c and i:

$$r_c = \frac{A\gamma_{\text{edge}}}{\rho_sK_{\text{tr}}}\frac{1}{i} \tag{22}*$$

By putting the condition

$$\bar{l} \geqq 2r_c$$

and introducing (19) for $\bar{l}$ they obtained an inequality which i should satisfy for growth to continue:

$$\bar{l} = K_1\left[\frac{r}{c_{i,b}}\right]^{1/2}i^{1/2} \geqq 2r_c = \frac{2A\gamma_{\text{edge}}}{\rho_sK_{\text{tr}}}\frac{1}{i} = \frac{K_2}{i} \tag{23}$$

then

$$i^{3/2} \geqq \frac{K_2}{K_1}\left[\frac{c_{i,b}}{r}\right]^{1/2}$$

or

$$i \geqq i_c = \left[\frac{K_2}{K_1}\right]^{2/3}\left[\frac{c_{i,b}}{r}\right]^{1/3} \tag{24}$$

* This relation is implicit in equation (8) of Ref. 50.

It is necessary to stress that, although the *physical model* presented by Price, Vermilyea, and Webb seems reasonable and it is probably the correct one in a number of cases (at least those involving deposition in the presence of organic additives), there are several points in the mathematical treatment of the model which should be re-examined. The most important of these concerns the use of the concept of critical radius and its relation to the overpotential. In an experiment involving deposition at a fixed overpotential, r_c represents a limiting radius of curvature below which no step can accept depositing particles. But when growth is actually taking place at a finite rate, part of the overpotential η_{tr} will be related to processes involved in the transfer of ions from the bulk of the solution to a point just in front of a growth step, prior to incorporation; the next step in the deposition process, incorporation at the growth site, requires the participation of an extra-overpotential, related to the need for the (curved) growth step to overcome the energy associated with its curvature. This "edge overpotential" will be given by an equation identical to (21):

$$\eta_{edge} = \frac{A\gamma_{edge}}{\rho_s \bar{r}_{st}} \qquad (25)*$$

where $\bar{r}_{st}$ is the *actual* radius of curvature of the growth steps (obviously an average value). In the presence of an adsorbed impurity, $\bar{r}_{st}$ will be related to its surface concentration:

$$\bar{r}_{st} = \frac{\bar{l}}{2} = \frac{1}{2}(N_0 \Gamma_i)^{-1/2}$$

and, therefore, to the current density [equation (19)]:

$$\bar{r}_{st} = \frac{K_1}{2}\left[\frac{r}{c_{i,b}}\right]^{1/2} i^{1/2}$$

which when substituted in (25) gives

* In this equation, the potential dependence of the interfacial tension, which would introduce a potential dependence in A has been neglected for simplicity.

$$\eta_{\text{edge}} = \frac{2A\gamma_{\text{edge}}}{\rho_s K_1}\left[\frac{c_{i,b}}{r}\right]^{1/2}i^{-1/2} \tag{26}*$$

Finally, the total overpotential will be given by the sum of η_{edge} and η_{tr}:

$$\eta_T = \frac{2A\gamma_{\text{edge}}}{\rho_s K_1}\left[\frac{c_{i,b}}{r}\right]^{1/2}i^{-1/2} + K_{\text{tr}}i \tag{27}$$

It is clear that the situation which makes the first and second terms in (27) equal† is a very atypical one. At any given applied overpotential, the steady-state current density will be such as to satisfy (27). There are also difficulties in attaching clear meaning to equation (22), interrelating a "radius of the critical nucleus" and the current density passing through the electrode, since the concept of critical radius is a limiting one corresponding to a "no growth" situation (i.e., $i = 0$).

It would be desirable to analyze the problem, both theoretically and experimentally, under potentiostatic conditions since in the galvanostatic case the growth of any one whisker is affected by the conditions of growth on the rest of the electrode, i.e., it is not possible to control the current (or current density) passing through each whisker, and the theoretical analysis of the galvanostatic case would involve parameters such as number of nuclei created in the initial potential peak, distribution of growing whiskers with respect to radii, etc.

Further, it is not always possible to neglect the concentration gradient of the metal ion in the solution when the current density at the whisker tip might be as large as 10 or 20 A/cm². In this situation it may be necessary to take into account the diffusional field created in the vicinity of the whisker.‡   A

---

* In fact, a minimum value of $\bar{r}_{\text{st}}$ is used here and, therefore, a maximum value of η_{edge} is obtained.

† This is the assumption implicit in the derivation of Price, Vermilyea, and Webb.

‡ If $D = 10^{-5}$ cm²/sec and $c_b = 6 \times 10^{-3}$ mole/liter, then for $r = 1\,\mu$, $i_{\text{lim}} = 60$ A/cm² while for $r = 10\,\mu$, $i_{\text{lim}} = 6$ A/cm².

strong effect of r on the current–potential relationship would then result in some cases, as opposed to equation (22).

Thus, the analysis of Barton and Bockris, which assigns a fundamental role to the enhanced rate of diffusion of ions toward the projecting tip of the growing needle (hence, introducing an r-dependence in the i–η relationship), appears more realistic for dendritic growth in pure solutions or melts. However, it is possible that, under heavy poisoning conditions, a mechanism qualitatively similar to that proposed by Price, Vermilyea, and Webb would become operative.

The primary effect of the rate of mass transport of the depositing ions on the mechanism of dendritic growth has recently been confirmed by T. B. Reddy for the galvanostatic growth of Ag dendrites from molten AgCl–LiCl–KCl mixtures.[164]

Finally, in comparing the experimental data with the predictions of the theory, Price, Vermilyea, and Webb found that the *actual* rate of growth of Ag whiskers seems to conform to the value predicted for the critical current density i_c, i.e., the growing crystal "adjusts itself" in such a way that the rate of growth corresponds *exactly* to the critical current density, rather than the latter being a lower limit for that rate. The theory does not interpret the mechanism by which this adjustment occurs.

Chapter 9

Mechanistic Aspects of Morphology

1. PRELIMINARY OBSERVATIONS

Morphological aspects of electrocrystallization are certainly the most examined aspects of the field. However, most observations are essentially *qualitative*. This situation originates from two distinct causes: (*a*) the emphasis on the technological aspect of electrodeposition results in the use of complex (i.e., technologically efficient) systems; the possibility of successfully interpreting those kinds of studies without fundamental background ones is very limited (e.g., the apparent, geometrical current density is usually recorded, and additives, the adsorption characteristics of which are not known, are employed, etc.). (*b*) The stress on the crystallographic aspects leads to detailed studies of the crystal forms obtained under different conditions rather than the (perhaps more necessary) kinetic observations that would throw light on the mechanistic aspects of crystal growth.

2. SOME FORMS OBSERVED

Although a very wide range of forms has been observed, most of them can, for single crystals, be typified by one of the following basic structures:

1. Pyramids, Fig. 21a (usually obtained at low current densities).[80,118,119]
2. Layers, Fig. 21b.[80,119,129]

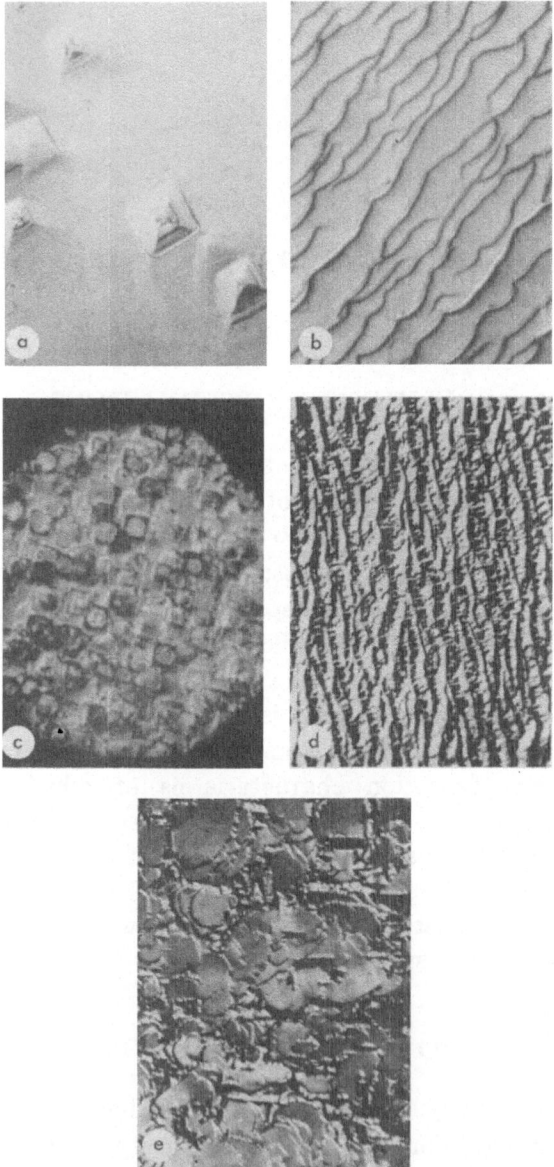

Fig. 21.   Commonly observed growth forms: (a) pyramidal growth; (b)
layer growth; (c) blocks; (d) ridges; (e) cubic layers. [(a) to (d):
Damjanovic, Paunovic, and Bockris[80]; (e): Seiter, Fischer, and
Albert[119]].

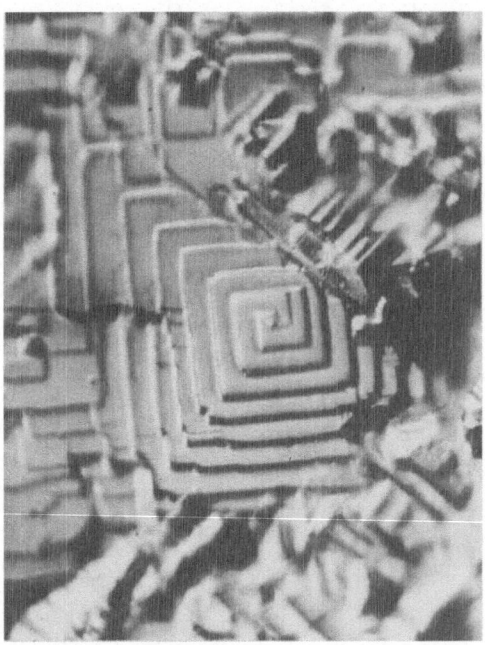

Fig. 22. Spiral growth in the electrocrystallization of Cu with pulsed current (Seiter and Fischer[118]).

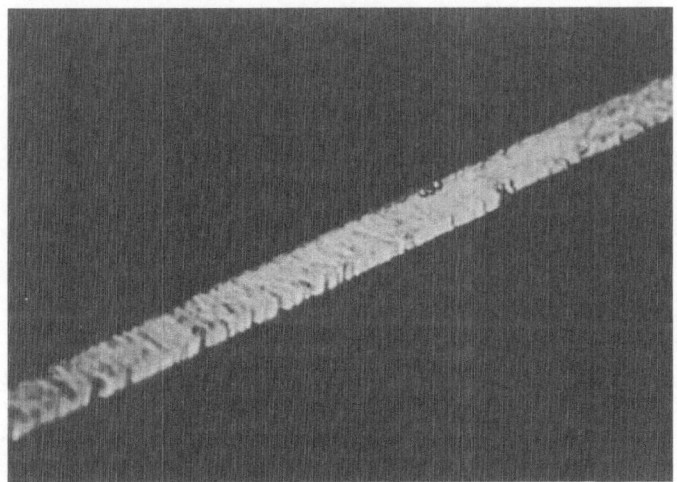

Fig. 23. Ag whisker grown electrolytically (Graf and Weser[163]).

Many other forms can be related to either one of these structures or intermediate ones: blocks (Fig. 21c) can be considered as truncated pyramids; ridges (Fig. 21d) as a special kind of layer growth; and cubic layers (Fig. 21e) as an intermediate structure between blocks and layers.

In addition, there are several less frequently observed growth structures:

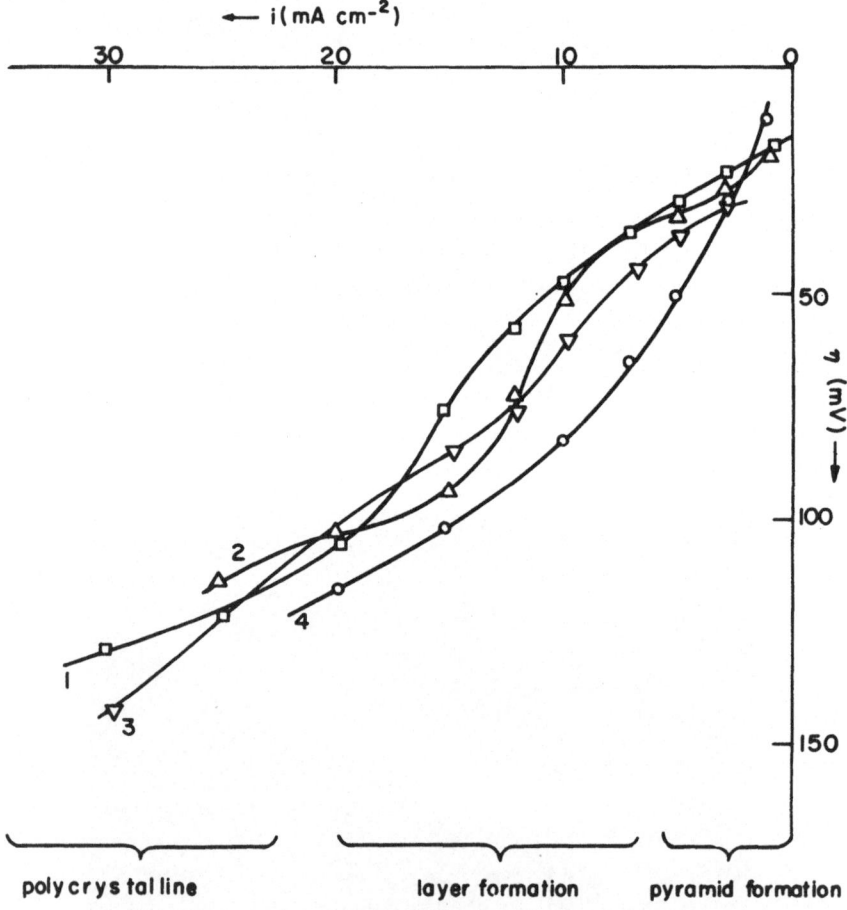

Fig. 24.   Correlation between polarization curves and growth form observed for Cu electrocrystallization. Curve 1: Seiter and Fischer; curve 2: Sroka and Fischer; curve 3: Shrier and Smith; curve 4: Mattson (Seiter and Fischer[118]).

1.   Spiral (Fig. 22).[118]
2.   Whiskers (Fig. 23).[120]
3.   Dendrites (Fig. 18).[121]

This classification is somewhat arbitrary, since it is based on the visible surface topography rather than on the sub-microscopic, elementary features.   Thus, pyramids might originate from elementary (invisible) spirals, whereas, the observed spirals are probably due to complex aggregates of elementary screw dislocations;[123] often the surface of the dendrites shows a layer structure; etc.   However, some support for the proposed classification arises from the experimentally established correlation between growth forms and overpotential (Fig. 24).[118]

Directly connected with morphological problems is the observation[80,127,169] that the activity of the surface depends strongly upon the growth forms which develop during the deposition process.   Thus, in experiments at constant current density, the overpotential changes markedly as the crystal grows.   This change can be related to the formation of new crystal faces on the growing surface.   Of course, this brings about a change in the real surface area and, hence, in the actual current density, but this effect is usually of secondary importance; a more profound change is that resulting from the difference in activity between the original surface and the newly developed one as it is reflected in a variation of the average exchange current density of the electrode.   Let $\{h_0 k_0 l_0\}$ be the original surface and $\{hkl\}$ the new crystal face developed on the surface.   The average exchange current density is given by[127,169]

$$i_{av}^0 = [i_{h_0 k_0 l_0}^0 (1 - \theta) + i_{hkl}^0 \theta]$$

where θ is the fraction of the surface occupied by the new face $\{hkl\}$, and if $i_{hkl}^0 \gg i_{h_0 k_0 l_0}^0$, even a small value of A will produce a large variation in i_{av}^0 with respect to the initial value.   Furthermore, the difference in activity will give rise to a nonuniform distribution of current density which will affect the growth process and, hence, the current distribution

at a later stage. Thus, the analysis of the potential dependence
of the growth forms, as well as the kinetic parameters and
activity of different crystal faces, can throw light on the
mechanism by which a crystal adopts one or another growth
habit.

3. THE POTENTIAL DEPENDENCE OF THE GROWTH FORM

The significance of the potential in any attempt to a
theoretical interpretation of the electrolytic crystal growth
process arises from a consideration of the following factors:

1. The adion concentration and local current density
distribution between growth steps are characteristic functions
of the overpotential.[106] Hence, it will be possible to relate the
latter to some basic questions such as the frequency of nucleation on the flat surfaces between steps or the possible build
up of local concentration gradients in the solution surrounding
a growth step.

2. The "activity" of the steps, i.e., the fraction of steps
which are active in the growth process, depends on the value
of r_c, the radius of the critical nucleus, which in turn is a
function of the overpotential (cf. Chapter 7).[45,107]

3. The strong dependence on potential of the surface
concentration of the different species present in solution,
particularly organic substances, will be of decisive importance
in determining the growth form at a given potential.*

4. Secondary effects, such as the potential dependence of
the interfacial tension,[122] or the potential dependence of the
charge on the adions and the energy of interaction of that
charge with the electrostatic field at the interface, must be
taken into account in a more advanced stage of the theory.

Thus, if a relation is to be found between crystal growth
and electrochemical parameters, the overpotential rather than
the current density must be considered the relevant controlling

---

* However, the essentially kinetic character of the adsorption process
during growth must be taken into account.

variable, although both are, of course, interrelated. These concepts have not yet been extensively applied, with the possible exception of the overpotential dependence of the supersaturation (σ), namely[13,42,45]

$$\sigma = \exp\left(-\frac{zF\eta}{RT}\right)$$

Although the principal determining factor in the variation of growth form for a given system may be the potential, the role of the substrate must not be neglected. Thus, the heat of activation for charge transfer is obviously a function of position, because the potential energy of the final state depends on the intermolecular potential in which the newly formed adion finds itself. Hence, although the overpotential (essentially a macroscopic quantity) is independent of position, the local current density may be strongly position dependent, thus affecting the morphological characteristics.

4. IMPURITY EFFECTS ON MORPHOLOGY

Impurity effects may be basic to all normal deposition mechanisms, in that the slightest amount of impurities deeply modifies both the morphological and kinetic aspects of electrolytic crystal growth. Direct effects were noted, e.g., by Economou, Fischer, and Trivich[123] and by Barnes, Storey, and Pick.[130] Correlations with amounts present in the solution were made by these co-workers. Later, Green, Swinkels, and Bockris[124] developed methods whereby the amount of adsorbed organic molecules could be measured as a function of potential on a solid surface. Extrapolation of the corresponding isotherms[125] made it possible for Damjanovic et al.[18,127] to relate the deposition forms to the adsorbed quantities:

Effects of n-decylamine during copper electrodeposition from $CuSO_4$-H_2SO_4 solutions, were observed on the distance between steps on a {100} face even when $\theta < 10^{-4}$. At $\theta > 0.03$, the crystal form was entirely different from that observed in highly purified solutions. For example, at $\theta < 10^{-2}$ (at concentrations of n-decylamine lower than 10^{-7} mole/liter) the crystal growth form was layers with a distance between steps of about 10^{-3} cm. At $\theta > 10^{-2}$, the deposit observed after the passage of 0.5 C/cm^2 is of the ridge type (Fig. 21d). If the distance between steps on a layer type of deposit is considered, it decreases, at the same current and thickness of the deposit, when n-decylamine is added to the solution. Pyramids on {100} and {111} faces tend to become more readily truncated (at lower current densities) and to transform to blocks (cf. Turner and Johnson[128]). In a series of photographs of the same area, a pyramid formed at an early stage of deposition (after 2 C/cm^2 has been deposited), is seen to become truncated at a later stage (~ 5 C/cm^2) of deposition, and eventually it transforms into a block (~ 8 C/cm^2) which grows sidewise into a large square layer.[80] A polycrystalline deposit, which resembles that formed from pure solutions, develops at current densities (~ 20 mA/cm^2) lower than that (~ 40 mA/cm^2) at which a similar crystal growth occurs in the absence of adsorbed n-decylamine.

When purification of the solution includes only crystallization and pre-electrolysis processes, the distance between steps on layer types of structures ({100} face) is shorter than

Table 5

Type of Deposit on {100} Face of Copper Single Crystals as a Function of Current Density in the Presence of n-Decylamine*

Current density (mA/cm^2)	Appearance of crystal growth at magnification 600 ×	
	In the presence of n-decylamine	Absence of n-decylamine
5	Layers (truncated pyramids)	Layers (pyramids), larger (up to 50%) distances between steps.
10	Layers+truncated pyramids	Layers+pyramids, larger distances between steps, low tendency to form truncated pyramids.
15	Layers+truncated pyramids+blocks	Layers, larger distances between steps, lower tendency to form truncated pyramids.
20	Polycrystalline	Layers, truncated pyramids, blocks.

* $\theta = 10^{-3}$ (Concentration 10^{-8} mole/liter).[80]

if activated alumina and charcoal are used in solution purification. The same trend is observable both at 5 and 10 mA/cm² (Table 5).

5. TEXTURE

The phenomenon of *texture* (the existence of metals with preferred crystal orientation) is observed in metals when they are subject to mechanical stress under certain conditions, and in electrodeposits in an advanced stage of polycrystalline growth. A more precise definition of texture can be given in terms of the concept of *degree of orientation*[8]: in a polycrystalline deposit the orientation in space of each grain can be specified by giving the angles formed by the crystallographic directions of the crystal and the axes of a reference system fixed with respect to the macroscopic substrate. If all three crystal axes in all the grains (which together constitute the deposit) are in fixed relationship to the substrate, then one obtains *three-degree orientation* which is none other than a *single crystal*. If only one axis in all the crystallites is fixed relative to the substrate, the other two axes being *randomly* disposed, then one obtains *one-degree orientation* or texture. If all three axes are randomly oriented with respect to the substrate in the ensemble of grains, then the polycrystal is said to exhibit *random orientation*.

From the electrocrystallization point of view, it is of interest to find the mechanism by which the crystallites growing on a randomly oriented substrate develop a one-degree orientation in a direction perpendicular to the substrate surface, as shown by electron diffration (Fig. 25).[8,126] The particular texture depends on electrochemical and crystallographic factors,[126] and the development of *facets* is observed; a theoretical approach capable of explaining these facts is then required.

The theory developed by A. K. Reddy[56] is based on the correlation between two aspects of growth, one microscopic and the other macroscopic.

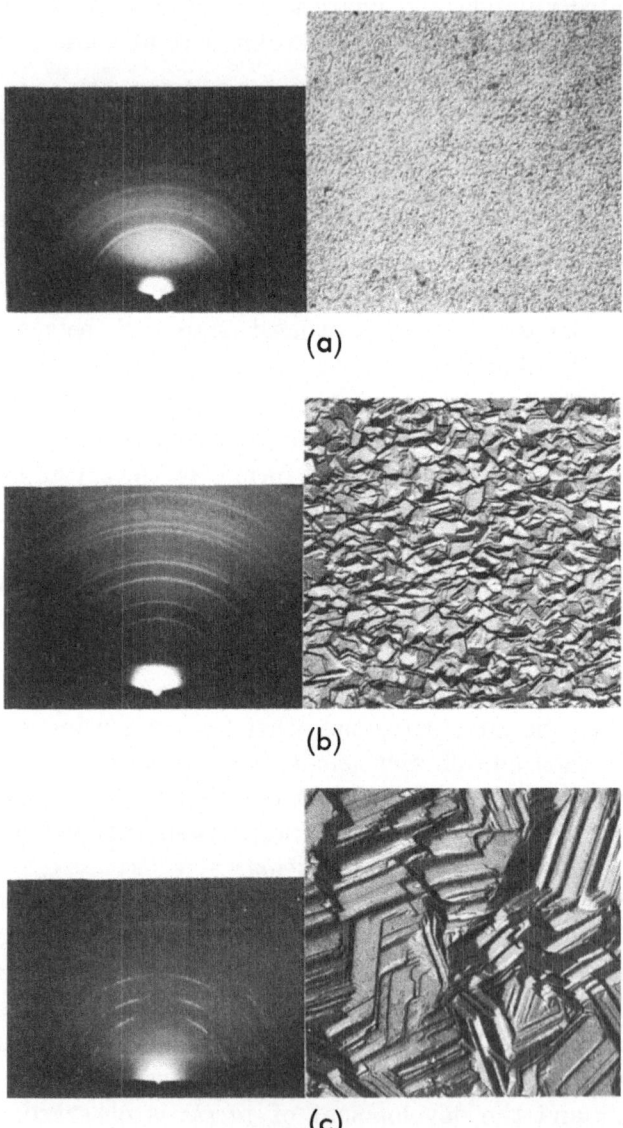

Fig. 25. Electron diffraction patterns and electromicrographs showing
development of preferred orientation in the electrocrystallization
of Zn at 40 mA/cm² ; (a) 700 Å deposit (10,000 ×), random orientation; (b) 6,300 Å deposit (5,000 ×), random orientation; (c)
54,000 Å deposit (5,000 ×), preferred orientation (Sato[126]).

1. *Microscopically*, a growing crystal has different kinetic parameters on different crystal faces. Thus, after a more or less long period of growth the crystal will develop well defined, usually low index, slow growing faces aligned according to fixed crystallographic planes. It is an essential point that these faces have different rates of growth.

2. *Macroscopically*, a crystal will have different rates of growth in a direction normal to the surface and in one parallel to it. This can be understood in terms of the influence of the electric field and concentration profile in the near vicinity of the growing substrate when an outward growth mode predominates.

3. A *correlation* then can be established between certain crystallographic axes and a set of axes referred to the substrate: The slowly growing faces will align themselves (see below) perpendicularly to the substrate surface leaving a given (fast growing) row of atoms perpendicular to the surface, i.e., producing a *preferred orientation*. The influence of additives, codeposition of H_2, etc., can then be explained[56,126] by considering the effect they will have on the kinetic parameters for different crystallographic faces; if that effect is large enough it will eventually result in different planes being the slowest growing, hence, in a different preferred orientation axis.

A different interpretation of the development of texture was offered by Pangarov,[57] the essential point being that, of the different crystallographic planes, the one which requires the least energy for nucleation will grow parallel to the substrate surface. The following assumptions are implicitly or explicitly made: (*a*) growth occurs by nucleation of successive layers, and (*b*) the substrate exerts no influence on the type of growth (i.e., it is "inert").

It is necessary to point out that, through its nonspecificity with respect to the molecular mechanism of growth, the theory proposed by A. K. Reddy offers a more general explanation for the formation of texture. In particular, its continuous character (*cf.* the stepwise character of Pangarov's mechanism) offers the possibility of understanding why a preferred orientation develops from a polycrystalline substrate and not from a single crystal: out of the ensemble of crystallites which constitute a polycrystalline substrate, a certain number of grains will have an orientation equal to (or close to) the preferred one. These crystallites will have a higher rate of outward growth than the others and will eventually, but not immediately, predominate in determining the surface topography.* Since each crystallite is effectively a single crystal, a nucleation theory would predict the development of preferred orientation on a single crystal substrate in much the same way as it does in the case of a polycrystalline one, in disagreement with experiment.

Finally, it is not likely that the growth process will involve a nucleation mechanism, especially at low overpotentials.[58,106]

---

* This explanation would imply a gradual coarsening of the deposit as the preferred orientation develops, and this actually has been observed.[126]

Chapter 10

Kinematics of Step Propagation

1. A NECESSARY METHODOLOGY

The movement of microsteps of a few angstroms in height across the surface of an electrode is not directly observable; however, these steps bunch together to form macrosteps and these are visible even to light microscopy, i.e., are several angstroms high. One can thus observe the movement of the crystal-forming entities, much as if one had a (much desired) electron microscope powerful enough to send a beam of electrons through a solution and record them after reflection. Rudimentary measurements of movements of such steps were published in 1932.[33] However, the advent of electron microscopy, with a magnification power 100 to 1000 times greater than that of light microscopy, drew attention away from such measurements, for it allowed pictures of much greater detail to be made of electrolytically grown crystals, *but only after they had been removed from the solution.* Measurement of the kinematics of the step growth were thus no longer made,* and the very important information obtainable from knowledge of the dependence of the rate of step movements as a function of concentration, potential, etc., was not measured for the next 30 years.

Another reason for this delay was lack of advance in the

---

* Kinematics of micro and macrosteps could be examined by means of electron microscopy if the growth were quenched, e.g., by the introduction during deposition of some suitable liquid. This laborious work has not yet been done.

technique of light microscopy. However, the availability since 1960 of the Nomarski techniques[75] (interference contrast microscopy and polarized interferometry, see Chapter 3) together with the introduction of improved techniques of solution control,[128,131] and electrochemical cells[76] made such observations possible.[18,80,127]

2. MICROSTEPS

Three sources of microsteps are available: (a) Those resulting from two-dimensional nucleation (Fig. 26a).* (b) Those resulting from emergent screw dislocations and other defects on the surface (Fig. 26b). (c) Those resulting from a misorientation in the cutting of the given single crystal (Fig. 26c).†

3. MACROSTEPS

When microsteps begin to move across a surface, they frequently come across impurity molecules‡ and stop moving, or move at a lower rate (Fig. 27, a and b). On top of the "first" layer-step, another step travels as far as the (blocked) first one. This process continues until about 1000 steps have formed on top of each other, i.e., a macrostep has been formed (Fig. 27c). Eventually, it appears as if the impurity molecules become buried by some unknown mechanism and the macrostep will move away (Fig. 27d).§

Two questions exist concerning the mechanism of blockage

* Not yet observed.

† These might be thought to be adventitious. In fact, it is not possible to cut a crystal plane to an accuracy greater than 1 or 2′; and hence such a cause of microsteps always exists in any cut crystal.

‡ This basic concept for blocking has never been subjected to direct proof.

§ This rather crude picture can be considerably improved if the concept of a "wave" of higher step density is utilized and if the model of a step or steps "stopped" by the presence of adsorbed impurities is replaced by a functional relationship between velocity of propagation of a microstep v_{st} and coverage θ such that $\partial v_{st}/\partial \theta < 0$.

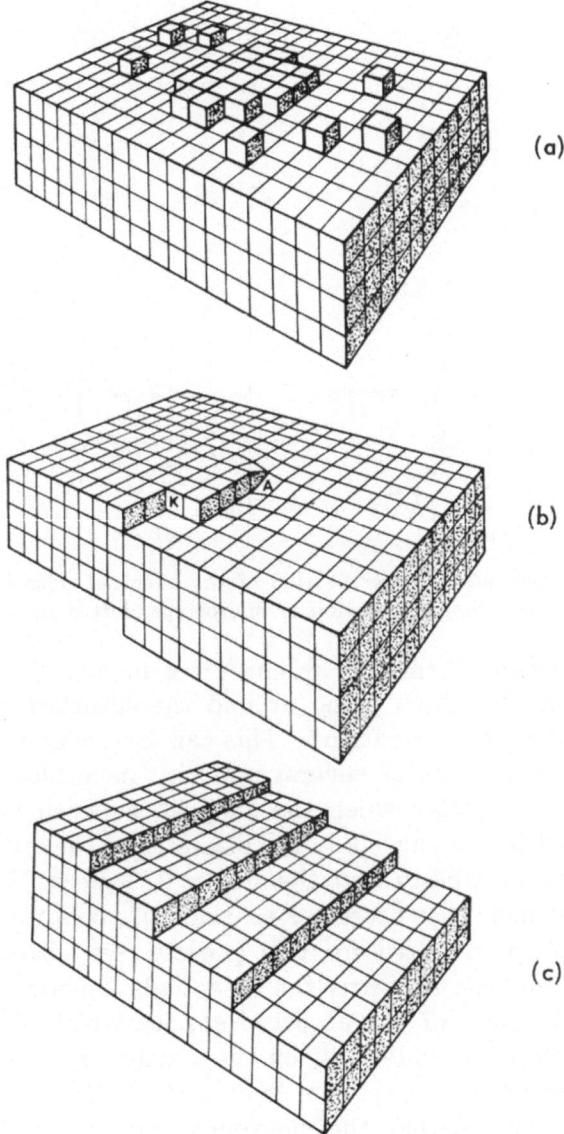

Fig. 26. Models of different sources of microsteps on a surface: (a) two-dimensional nucleus; (b) emergent screw dislocation (A) with kink site K; (c) misorientation of the surface with respect to the ideal low-index plane.

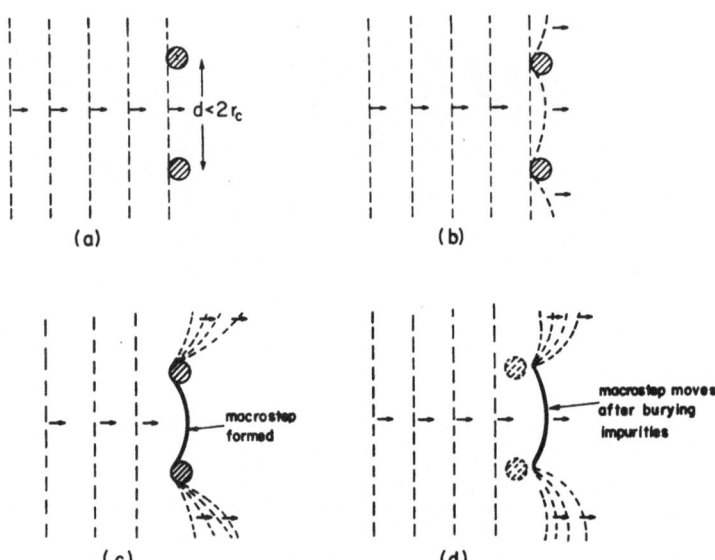

Fig. 27.   Schematic representation of the effect of adsorbed impurities
           upon the propagation of microsteps (dotted lines).

and macrostep formation arising from impurities. What prevents the step from going around the adsorbed molecule and
continuing its propagation? This can be understood in terms
of the joint effect of several impurity molecules adsorbed at
the edge (Fig. 27a); when the distance between two adsorbed
molecules is comparable to the diameter of the critical nucleus
($2r_c$), the probability that the step will "squeeze" between two
adsorbed molecules decreases considerably because of the large
edge energy required for that process (see Chapter 7). The
second question concerns the essentially kinetic character of
the equilibrium of the adsorbed species which requires that
an adsorbed molecule will, on the average, return to the solu-
tion after a time proportional to $\exp{(\Delta G^{0\ddagger}_{\mathrm{desorp}}/RT)}$. However, it
can be expected that the molecules adsorbed at an edge will
have a greater energy of adsorption than those adsorbed on
a flat surface (because of the greater coordination with surface
atoms) with the resulting increase in the free energy of
activation for desorption.

4. FACTS OF MACROSTEP MOVEMENTS

The upward growth of steps* can be followed by interferometric measurements, and the lateral one can be followed
by interference contrast microscopy.[75,80] Steps all propagate
in the same direction, and this does not depend upon the
direction of flow of the solution over the electrode on which
deposition is taking place.

The distance between the steps *increases* as the deposit
grows thicker (Fig. 28).[80] Correspondingly, the height of the
steps increases with the total thickness of deposit (i.e., when
the whole deposit is increasing in thickness, the height of one
step over the one beneath it increases).

* By this is meant "the variation with time of the step height"
since only differences in height can be recorded by the interferometric
technique.

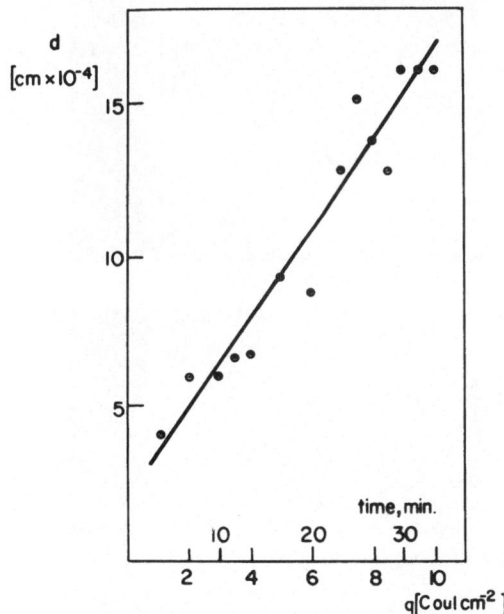

Fig. 28.   Variation of average distance between macrosteps with thickness of deposit for Cu electrocrystallization.

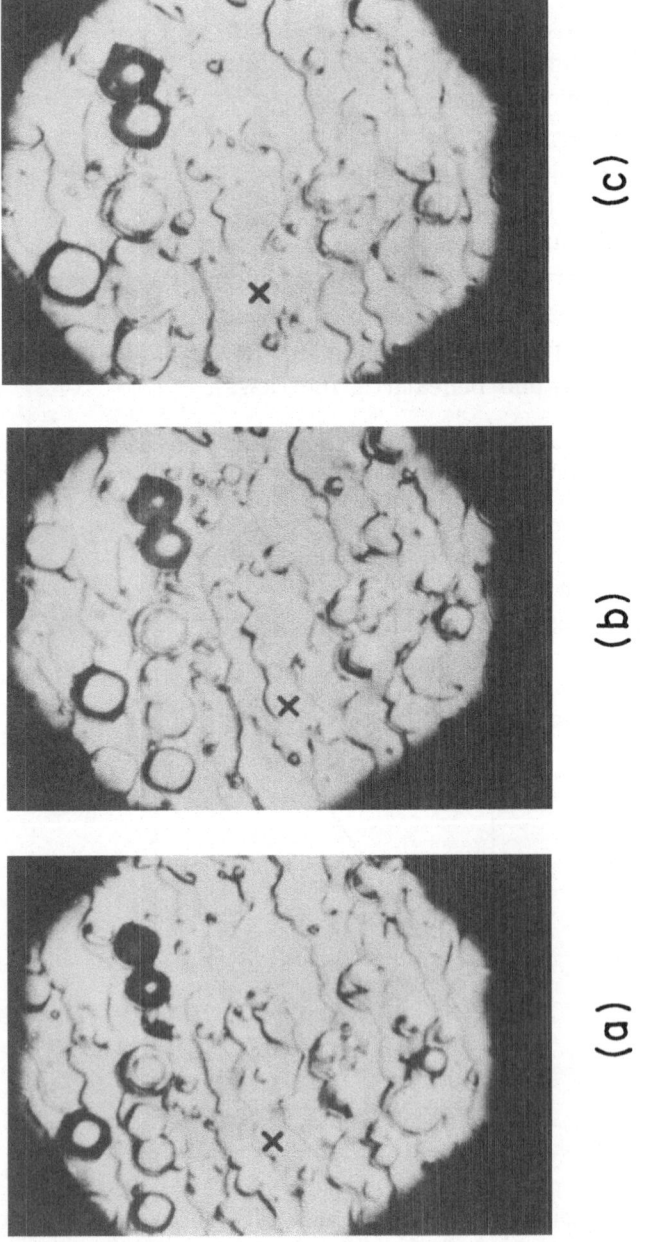

Fig. 29.   Fading out of macrostep (×), as observed *in situ* during electrodeposition of Cu: (a) 8 C/cm²; (b) 9 C/cm²; (c) 10 C/cm². (Damjanovic, Paunovic, and Bockris[80]).

The velocity of the lateral movement of macrosteps is about 2×10^{-6} cm/sec at a current density of 5×10^{-3} A/cm^2 for deposition of Cu on Cu from acidified CuSO$_4$ solutions. The amount of material needed to explain the advance of the macrosteps is only about one tenth of the total amount deposited, thus indicating growth of the crystal *between* macrosteps.

A surprising phenomenon is the observation of the *fading* of steps (Fig. 29).

5. DETERMINATION OF THE ORIGIN OF MICROSTEPS

The facts cited concerning the direction of propagation of the macrosteps suggest that the microsteps which give rise to macrosteps do not originate from emergent screw dislocations but are introduced by cutting the crystals not exactly parallel to the given plane. This hypothesis, first suggested by Howes,[132] was tested by Damjanovic *et al.* by the procedure of cutting the crystal at different angles with respect to the crystal axis. The direction of the macrostep propagation corresponded to that which the microsteps, created by the different directions of cutting the crystal, would have.

6. THE INSTABILITY OF BUNCHES OF MICROSTEPS IN THE ABSENCE OF IMPURITIES

The shape of a given crystal surface can be described in terms of the density of microsteps (number of microsteps per unit length) and its variation with position, i.e., the function $\rho_{st}(x)$.

A smooth surface will be characterized by an almost constant value of the density of microsteps: $\rho_{st}(x) = \rho_{st,0}$. A stepped surface, that is, one exhibiting large, visible macrosteps, will correspond to a density profile with large peaks indicating the position of the macrosteps (Fig. 30).

Growth of a crystal is often accompanied by a transformation of the surface from a smooth into a stepped form, and

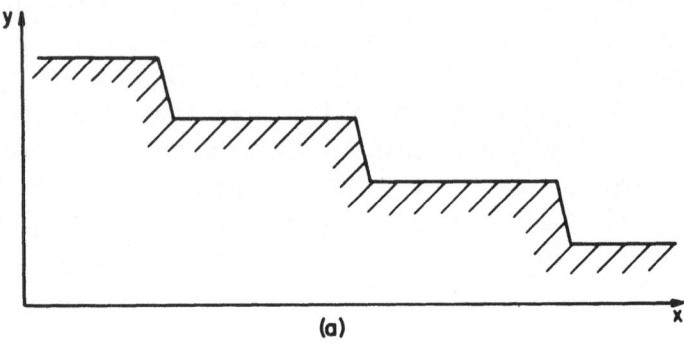

(a)

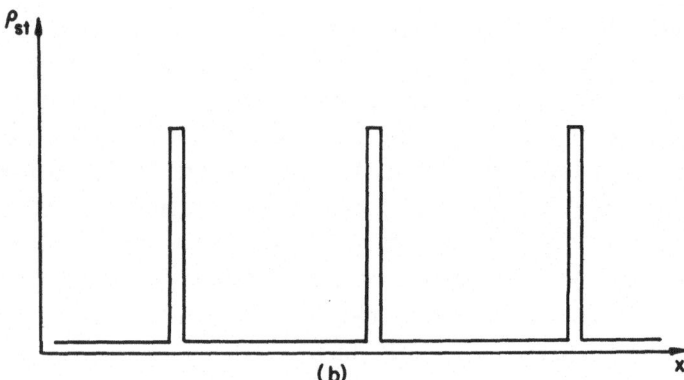

(b)

Fig. 30. (a): Surface with macrosteps. (b): The corresponding distribution of density of microsteps.

the kinematic analysis of the problem,[19,20] which attempts to give an interpretation of the mechanism for that transformation, is based on the consideration of the rate of propagation of the microsteps and the dependence of this upon the proximity of other microsteps.

However, this kinematic theory fails to provide an explanation for a process of spontaneous bunching on an initially smooth surface. Actually, the theory predicts that *in the absence of adsorbed impurities* any initially existing bunch (incipient bunch) will gradually become smoother, until it fades out. In the course of the process, the incipient bunch may develop a sharp edge before fading out.

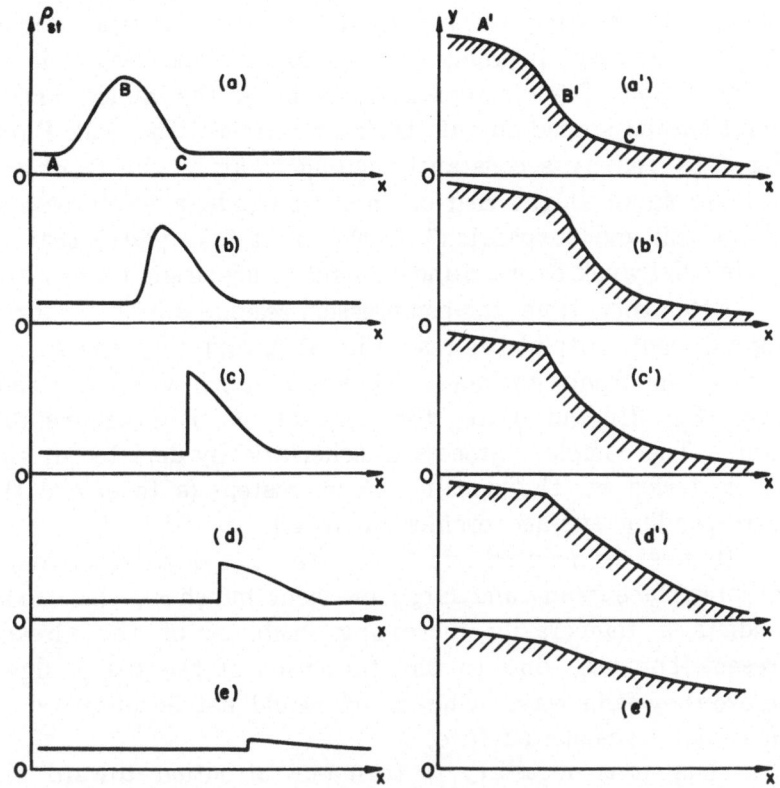

Fig. 31. Progress of an initially symmetric bunch of microsteps (a and
a') according to the kinematic theory; (a) to (e): variation with
time of the distribution of density of microsteps; (a') to (e'):
corresponding variation of surface profile.

Consider an incipient bunch described by the density of
microsteps distribution of Fig. 31a (arising, e.g., from a slight
and sporadic imprecision in the cutting of the crystal). It is
reasonable to assume that the velocity of propagation of
microsteps decreases with increasing density of microsteps,
because as ρ_{st} increases, less depositing ions will be available
for each microstep.* Thus, in the tail region AB (Fig. 31a),

---

* The model to be discussed is strictly correct only if surface diffusion
is in partial control of the rate of deposition, so that v_{st} does not vary
exactly as ρ_{st}^{-1} (pure charge-transfer control), but exhibits a slower
variation with ρ_{st}.

v'_{st} increases in going from A to B, while in the front region BC, v_{st} decreases in going from B to C. This results in an accumulation of microsteps at the rear of the bunch and a simultaneous spread-out at its front. Hence, the peak B (*cf.* Fig. 31a and b) is constantly fading away at the front and forming again at the rear or, in other words, a "compression" of the tail and "expansion" of the front take place; that is, the bunch travels forward and changes its shape simultaneously.

After some time, the compression reaches a limit by forming a discontinuity at the rear (Fig. 31c), and only the expansion of the front continues. Thus, the bunch will finally fade away (Fig. 31d and e), i.e., the incipient bunch is an unstable form. The complete process is schematically depicted in Fig. 31 in terms of the density of microsteps (a to e) and the corresponding surface profiles (a' to e').

It must be pointed out that in the case of a rather rough initial surface (many and large incipient bunches), this model predicts a temporarily increasing visibility of the already present bunches, due to the formation of the sharp edges, before they fade way. This trend would not be observed on an initially smooth surface.

Thus, it is necessary to turn our attention toward the influence of adsorbed impurities upon the mode of propagation of microsteps.

7. IMPURITY ADSORPTION THEORY OF BUNCHING

Frank[19] has qualitatively analyzed the effect of time-dependent adsorption of impurities upon the stability of a smooth surface. Consider a uniform distribution of microsteps as shown in Fig. 32, i. Let step D undergo a temporary decrease in its velocity of propagation, due, for example, to a fluctuation. The length of terrace d will thus increase, and the length of terrace c will decrease (Fig. 32, ii). Hence, the lifetime of the surface which the microstep C meets in its propagation will be reduced, and that of the surface met by step D will be increased with respect to that of the average

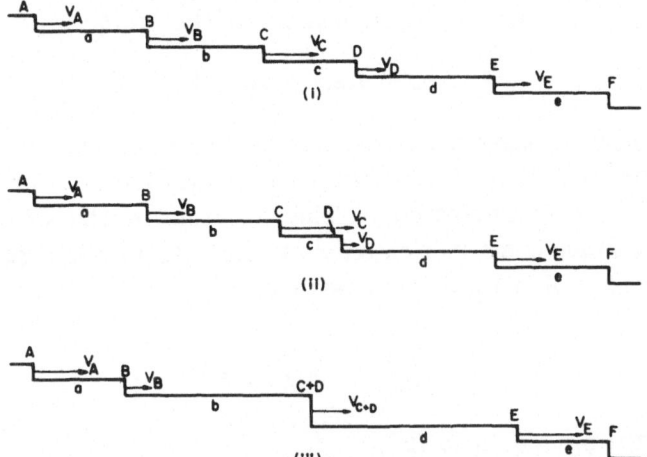

Fig. 32.   Schematic model for pairwise bunching of microsteps in the presence of impurities: (i) uniform distribution of microsteps on the surface (ii) a perturbation on step D is reflected on step C (iii) steps C and D collapse into a double step and the perturbation propagates to steps B and E.

terrace.   If it is assumed that the adsorption of the impurities is not at equilibrium, this is equivalent to saying that C will find a lower coverage and D a higher one, which in turn will result in C increasing its rate of propagation and D decreasing it further (indicated by the length of the V vectors in Fig. 32, ii) until C and D will eventually collapse to form a double step (Fig. 32, iii).   Simultaneously, the width of terrace b has increased, thus slowing down the step B, and step E has increased its speed slightly (due to the reduced competition with D).   That is, the process of bunching of microsteps in pairs is a spontaneous one or, in other words, the even distribution of microsteps is an unstable situation.   This pairwise bunching can continue, now involving the double steps, so that eventually steps high enough to be visible in interferometry will be formed on the surface.

Mullins and Hirth have given a general mathematical treatment of this situation.[170]   In essence the theory assumes that the rate of propagation (v_n) of a given microstep in a train is a function only of the distance (ε_n) to the microstep ahead and (ε_{n-1}) to the microstep behind in the train:

$$v_n = f(\varepsilon_n) + f(\varepsilon_{n-1}) \tag{28}$$

Furthermore, they assumed that the function $f(\varepsilon)$ would be the same whether it referred to the terrace ahead or to that behind the nth microstep.* The variation of the width of a given terrace is a function only of the rate of propagation of the steps which limit that terrace:

$$\frac{d\varepsilon_n}{dt} = v_{n+1} - v_n \tag{29}$$

Substituting (28) into (29),

$$\frac{d\varepsilon_n}{dt} = f(\varepsilon_{n+1}) - f(\varepsilon_{n-1}) \tag{30}$$

This gives a set of simultaneous differential equations which can be solved analytically or numerically in any particular case for which the function $f(\varepsilon)$ is known.

This approach has been applied by Hulett and Young[172] in the case of anodic dissolution of Cu in HCl and HBr solutions. They adopted a function $f(\varepsilon)$ experimentally determined,[171] and solved numerically the set differential equations with $n = 100$ and cyclic boundary conditions. A good agreement with experiment was found, but no account was given for the dependence of $f(\varepsilon)$ on macroscopic variables (concentration, current density), and neither was that function explained in terms of a detailed atomistic model.

8. MECHANISM OF THE PROPAGATION AND FADING OF MACROSTEPS

The increase of the interstep distance with rate of deposition can be rationalized by the fact that the microsteps

* This makes difficult the introduction of time dependent adsorption of impurities in the present model, since the effect of ε_n on v_n would be basically different from its effect on v_{n+1}.[19]

can grow twice as far per unit time at $10 \, mA/cm^2$ as at $5 \, mA/cm^2$. A given point on the fresh surface will, hence, be exposed to the solution for a shorter time at higher than at low current density, the steady state (nonequilibrium) surface concentration of impurities will be smaller,* and hence the probability of the initiation of bunching is reduced.

The fact that only a small fraction of the total material deposits on the macrosteps implies, of course, that most of the material goes to the growth of layers between macrosteps, and that this material grows on steps invisible at interference

* The assumption of impurities being buried by the advancing layers is implicit.

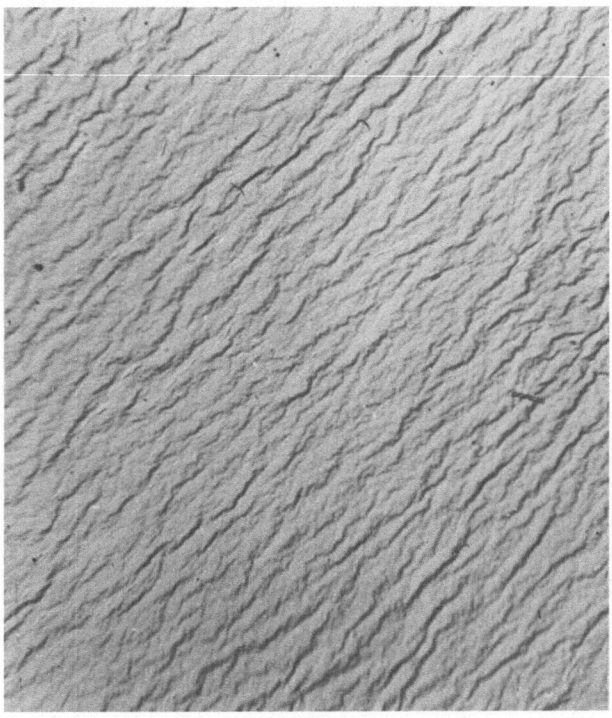

Fig. 33. Electromicrograph of Cu deposit (20,000 ×) showing microsteps in the "flat" region between macrosteps of a layer-type deposit (*cf.* Fig. 21b). (Paunovic, unpublished).

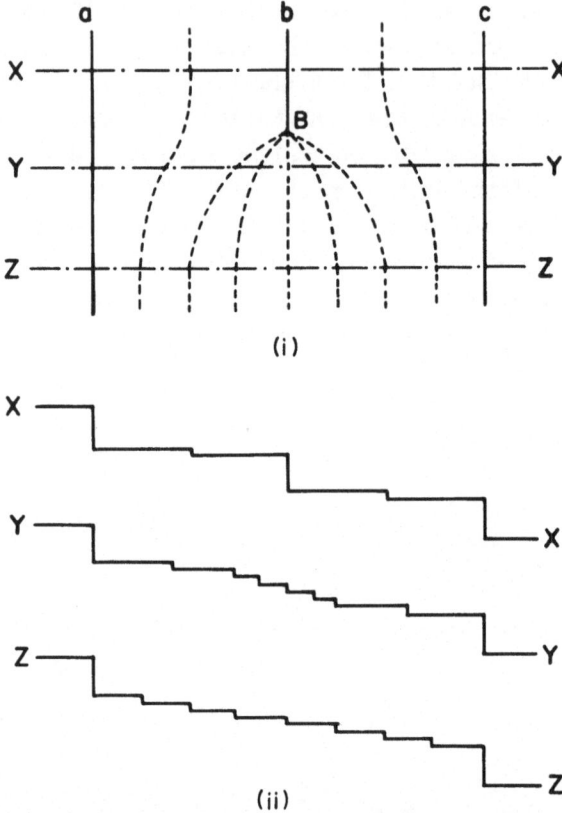

Fig. 34.   Model for fading out of a macrostep from its end. (i) top view
          showing macrosteps (a, b, and c) and microsteps (dotted lines);
          (ii) cross sections of surface shown in (i).

contrast microscopy (i.e., $h < 100$ Å).   This was confirmed in
one case[80] by electron microscopic observation.   Here (Fig. 33)
one can observe the microsteps.*

  The fact that the interstep distance increases with time
means either that one step moves to catch up another, or that
one step fades away.   The latter was more often observed
than the former by Damjanovic *et al.*

  Fading out of macrosteps can be visualized by the model
in Fig. 34.   Macrosteps a, b, and c are cross sectioned at X–X,

---

  * These, of course, are not monatomic steps.

Y-Y and Z-Z. Broken lines represent smaller, perhaps monatomic, steps. In the vicinity of the emerging point B of step b the density of micro (monatomic) steps is much higher than at any other position between the macrosteps. This is represented by the cross section Y-Y. Fading out of a step (step b) from its end (B) can be understood as a decomposition of the macrostep into smaller microsteps. Eventually, these microsteps distribute evenly over the surface between the adjacent macrosteps (a and c). After a sufficiently long time, the cross section X-X would appear like the cross section Y-Y, whereas Y-Y would look like Z-Z. The density of microsteps may, however, be higher to the right of the fading step, since its fading may involve the sending away of microsteps faster than the receiving of them (i.e., "bulk-fading" superimposed with "end-fading"). The cause for the spreading of microsteps may lie in the desorption of impurities previously adsorbed at the edge.

Chapter 11

Effect of Adsorbed Substances
upon Electrocrystallization

1. FACTS

It is well known[55] (and wide use of this knowledge is made in practice) that the addition of certain organic substances affects to a very large extent the physical properties, and the conditions of formation, of electrodeposits. An important characteristic of these effects is the low threshold for their action; amounts of substance in solution in concentrations as low as 10^{-12} mole/cm^3 have been found[80] to produce significant changes in the morphology of an electrolytically growing crystal. Indeed, there are indications[19] that at least some of the characteristics observed in crystals grown from "pure" solutions are in fact determined by the presence of very small amounts of adventitious impurities.

From a phenomenological point of view two kinds of effects produced by the additive are discernible: (a) morphological ones, i.e., those resulting in changes in the microscopic (or even visible) appearance of the surface, as well as in the structural properties of the deposit (grain size, orientation, stress, dislocations, etc.); (b) electrochemical ones, resulting in changes in the η-i relationship for the particular electrode reaction (e.g., variations in i_0 and b for a Tafel relation, or disappearance of a Tafel behavior). Both aspects are, of course, interrelated and are simply descriptions which must be interpreted in terms of modifications produced by the addition agent in the mechanism of the electrocrystallization process. Hence, the systematic study of the effect of controlled amounts of impurities in solution upon the characteristics of crystal growth, besides yielding valuable technological information, would greatly help in the understanding of the basic mechanism of that process.

Although the accumulation of inhibitor* in a layer of solution adjacent to the electrode appears capable of affecting the deposition process to some extent (electrolyte film inhibition[65]), the basic action of the addition agents seems to arise almost invariably from their adsorption on the electrode surface and, hence, modification of the conditions of the steps of the crystallization process following that of mass transport in the solution. Thus, a clear correlation has been established[133] between adsorption of some organic compounds (as measured by the decrease in double layer capacitance) and the modification of the electrochemical kinetic parameters for the Ni deposition and dissolution reactions on the same electrode.

The repeated observation[65] of the decreasing effect of an uncharged addition agent as the electrode potential departs from the value corresponding to the point of zero charge is very significant (cj. relation between coverage by organic molecules and charge on the electrode[103]). The reduction in grain size of polycrystalline deposits upon addition of inhibitors affords another example of the localization of the inhibitor effect at one of the later stages of the deposition process. Grain refinement can be qualitatively interpreted in terms of the increased frequency of nucleation, which results from the higher concentration of adions on the surface; an increase in the adion concentration can only be explained by assuming that the inhibitor hinders a step for which the adions constitute a reactant, i.e., surface diffusion or incorporation at a growth

* The term "inhibitor" arises because adsorbed substances usually *decrease* current at a given overpotential, i.e., they have a *negative* catalytic activity with respect to electrocrystallization. Here, however, it is used in the wider sense of "substances which affect the electrode-position and growth process," i.e., inhibit the "free mode" of growth and impose a different one.

site. This in turn requires that the inhibitor be adsorbed on the surface in order to be effective.

A different question concerns the fate of the inhibitor once it is adsorbed on the surface; no general answer can be given. In some cases, burying of the adsorbed particles into the growing metal seems to occur; in other cases, the adsorbed substance is electrochemically reduced to less adsorbable species which return to the solution.[149]

2. ADSORPTION OF INHIBITORS AT ELECTRIFIED INTERFACES

One of the difficulties of inhibitor studies is the lack of information concerning the adsorption of molecules on electrode surfaces and its dependence upon the electrode potential and the concentration of the substance in the solution. Reliable data for adsorption of inorganic ions on Hg have been available since the 1950's[134-137]; in the case of organic substances similar information was obtained around 1960[112,138,139] (but see Frumkin[140]). However, from the point of view of electrocrystallization, the adsorption on liquid electrodes is relatively unimportant, and the corresponding information for the solid metal electrode is scarce. Recently, methods have been developed[84,124,141] (*cf.* Ref. 142) whereby a *direct* measurement of the amount adsorbed on the surface is obtained under equilibrium conditions (i.e., without removing the electrode from the solution or altering its polarization state). These methods have not been extensively applied as yet, but they provide a valuable tool to obtain accurate adsorption data.[84,125,141,143,144] Although insufficient data exist to make many summarizing statements concerning the relation of the degree of adsorption to molecular structure for the solid metals, some generalizations can be made for adsorption of organic molecules on Hg, and this probably carries over well to solid metals. Thus:

1. The coverage–potential relation is approximately parabolic (Fig. 35) except: (*a*) if at the upper regions saturation is reached; (*b*) for the aromatic cations (e.g., $C_6H_5NH_3^+$), whose

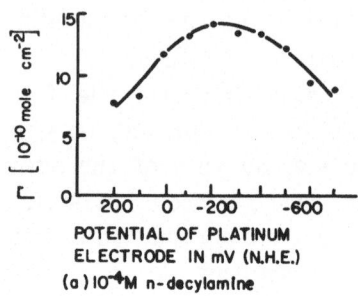

(a) 10^{-4}M n-decylamine

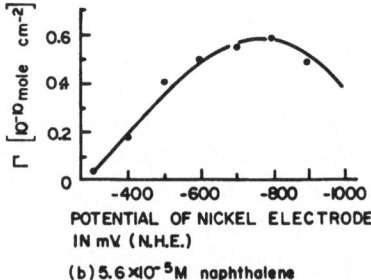

(b) 5.6×10^{-5}M naphthalene

Fig. 35.   Adsorption of organic substances on solid electrodes as a function
of potential: (a) n-decylamine ($c = 10^{-7}$ mole/cm³) on Pt (Bockris
and Swinkels[125]); (b) naphthalene ($c = 5.6 \times 10^{-8}$ mole/cm³) on Ni
(Bockris, Green, and Swinkels[143]).

Table 6

**Intrinsic Standard Free Energy of Adsorption for
Different Organic Groups[112]**

Group	(ΔG^0_{ads}) intrinsic (kcal/mole)
Butyl	-6.4
Phenyl	-8.9
Naphthyl	-12.4

π-electron ring may interact with the metal surface in the anodic
region of the coverage–voltage curve, while they are adsorbed
as regular cations in the cathodic side.   In such systems (e.g.,
that of aniline hydrochloride in acid solution in contact with
Hg),[138,177] there is little (30–60%) change in the degree of
coverage with potential over the whole potential range.

2.   The coverage–potential curve tends to be symmetrical

with respect to the point of zero charge, particularly in the case of uncharged aliphatic compounds.[125]

3. As the aromaticity of the molecule increases, the adsorbability increases greatly, as shown in Table 6, where the intrinsic standard free energy of adsorption for three different groups of compounds is given. The use of intrinsic values[112] is equivalent to two hypothetical corrections: (a) the free energy related to desorption of water (upon adsorption of one mole of organic) is subtracted; (b) the intrinsic values correspond to a hypothetical member of the group such that no free energy change occurs upon removal of one mole of organic from the solution (zero standard free energy of solution). Secondary effects arise for substituents. The difference in adsorbability between different substances corresponds to more than four powers of ten. Thus, it is possible to choose a compound to adsorb with any coverage at any potential.*

Essentially, similar relations exist for the plots of coverage, θ, or surface excess, Γ, as a function of potential, for different organic molecules (Fig. 35). The explanation of the parabolic dependence arises[103] because of the competition for the sites on the surface between water and the organic

Table 7

Simple Model for Potential Dependence of Adsorption of Neutral Organic Substances

Charge on the metal (q^M)	Orientation of adsorbed water dipoles	Adsorbability of water on clean metal (metal-water bond strength)	Adsorbability of organic on clean metal (metal-organic bond strength)	Net adsorbability of organic ($-\Delta G_{ads}^0$)
Large and positive	↓ ⧸⧸⧸⧸⧸⧸⧸⧸⧸⧸⧸	High	Constant	Low
Zero	Approximately random	Low	Constant	High
Large and negative	↑ ⧸⧸⧸⧸⧸⧸⧸⧸⧸⧸⧸	High	Constant	Low

* This is a useful result, particularly with respect to corrosion inhibition studies.

molecule. The water dipoles are oriented (see Table 7 and Fig. 37) predominantly with their oxygen atom toward the metal when the latter has a net positive charge; and when the sign of the charge on the metal reverses, the dominant orientation of the water molecule reverses, so that, as one departs from the point of zero charge on the metal, there is an increasingly strong attachment of water molecules to the electrode. The organic dipoles usually arrange themselves with their dipoles right outside the Gouy layer at the electrode, and hence they are unaffected by the changing electric field. Therefore, they adsorb to the greatest extent when the water molecules are least attached to the metal, i.e., in the first approximation of the model, at the point of zero charge (see Table 7). This potential dependence of adsorption is obviously of the greatest importance in respect to the potential dependence of inhibitor effects in deposition.

However, although the simple model (Table 7) indicated that the organic molecules will adsorb in a way symmetrical to the point of zero charge (Fig. 36), the available data

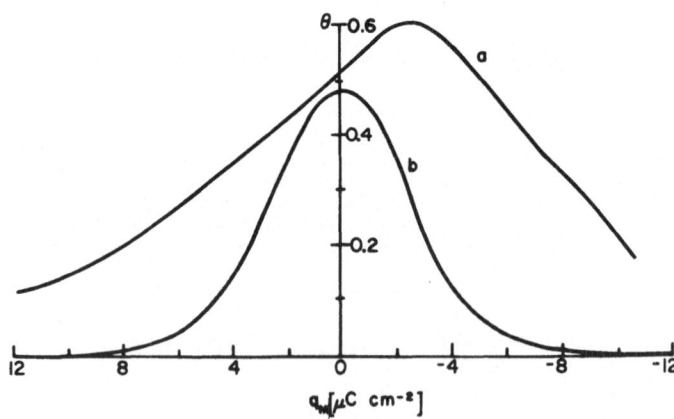

Fig. 36.   Comparison of the experimental adsorption curve (a) for 0.1 N butanol with a theoretical curve (b), without taking into account lateral interaction (Bockris, Devanathan, and Müller[103]).

Table 8

Potential Difference Between the Point of Zero Charge and the
Adsorption Maximum of Naphthalene and n-Decylamine for $\theta \to 0$

Metal	$\Delta E = E_{max} - E_{p.z.c.}$ (volts)	
	Naphthalene	n-decylamine
Ni	-0.3	-0.2
Fe	-0.2	-0.2
Cu	-0.7	-0.7
Pt (alk.)	0.0 to -0.1	0.0 to -0.1
Pt (acid)	-0.2	—

suggests that not only are the maxima displaced from this
point but also that the displacement (i.e., the potential difference between the potential of maximum adsorption of the
organic molecule and that of the zero charge) is dependent
upon the metal (Table 8).[125,143]

The interpretation, in principle, of a shift requires a
small modification of the theory of adsorption of organic
molecules on an electrode. Thus, a theoretical equality $E_{max} =$
$E_{p.z.c.}$ obtains only upon the assumption that there is no
chemisorption and that the strength of attachment of the
water molecule is equal for both orientiatons at $q^M = 0$.
If this is not so, it turns out[103,143] that there will still be
symmetry around some potential but that the potential concerned will be shifted from that of the point of zero charge
by an amount which corresponds to the energy difference in
the attachment of the water molecule to the electrode in the
"O up" and "O down" positions* (Fig. 37). These largely
electrostatic considerations of the adsorption of water and the
potential dependence of inhibitor attachments to metal surfaces
certainly do not indicate an advanced model. For example,
there is also evidence (cf. Ref. 138) that a specific interaction
occurs between the metal and the π-ring of the aromatic
molecule.[143]

The description given here of inhibitor adsorption outlines

* "O down" means "oxygen toward metal."

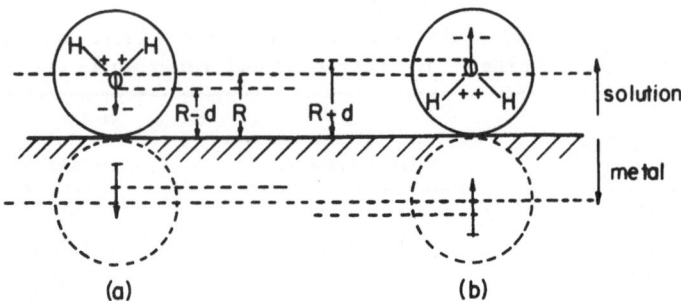

Fig. 37.   Difference in the image interaction for the two principal positions of water dipoles at the interface (Bockris, Devanathan,
and Müller [103]).

the situation corresponding to the relatively homogeneous
adsorption which applies at $\theta > 0.1$.   It is known that there
is an effect of inhibitors at a coverage much less than this,
which would indicate heterogeneous adsorption at growth sites.
It is probable that the heat of adsorption of an organic molecule, e.g., at a kink site, can be much greater than it is on
the planar surface. However, water molecules are also adsorbed
at growth sites and there is no knowledge available at the
present time which suggests that there will be a different
symmetry of adsorption of the water molecule on the growth
sites.   Therefore, the potential dependence of the adsorption
of water molecules should be approximately as great at kinks as
it is at the planar sites, to which most of the above discussion
applies.   In other words, the heterogeneity of the surface
would tend to be masked by the equalizing effect of the
adsorbed water molecules, at least with respect to the potential
dependence of adsorption.   Whether at each potential the distribution of adsorbed substance with respect to surface sites
will be homogeneous or heterogeneous will depend upon the
variation of $[\Delta G_v^0(\text{org}) - n\Delta G_v^0(\text{w})]$ from one point to another
on the surface [$\Delta G_v^0(\text{X})$ is the standard free energy of adsorption
of X from the pure vapor phase onto the clean metal surface].
It is reasonable to expect both $\Delta G_v^0(\text{org})$ and $\Delta G_v^0(\text{w})$ to become more negative in going from a planar to a kink site,

but whether the *difference* between ΔG_v^0 (org) and ΔG_v^0 (w) becomes more negative or does not change will depend on the particular case. Therefore, the heterogeneity of the adsorption cannot be predicted in a general sense.*

3. THERMODYNAMICS AND KINETICS OF ADSORPTION

(i) Dependence of the Equilibrium Adsorption on the Metal for a Given Inhibitor

The extent to which a substance will be adsorbed at a metal–solution interface depends upon the value of the standard free energy of adsorption. Assuming a Langmuir isotherm:†

$$\frac{\theta}{(1 - \theta)^n} = kc_b \tag{31}$$

with

$$k = \frac{\exp\left(-\Delta G_{ads}^0/RT\right)}{0.0555} \quad \text{cm}^3/\text{mole} \tag{32}$$

where the factor 1/0.0555 arises from taking as standard state unit mole fractions, while c_b (the concentration of organic substance in solution) is in moles/cm³. Using simple thermodynamic cycles, it can be shown[112,125] that

$$\Delta G_{ads}^0 = U + \Delta G_v^0 (\text{org}) - n\Delta G_v^0 (\text{w}) \tag{33}$$

where U is a constant for each organic substance, reflecting the interactions in the solution phase (water–water, organic–organic, water–organic), ΔG_v^0 (org) is the standard free energy of adsorption of the organic from the vapor phase onto a clean (water-free) metal surface, ΔG_v^0 (w) is the corresponding magnitude for water, and n is the number of water molecules displaced by the organic. Experimental determinations of ΔG_{ads}^0, as well as theoretical calculations of the second and

---

* Certain indications of heterogeneous adsorption arise precisely from the effect of very low concentrations of organic substances upon crystal growth.

† Modified to account for n-site adsorption.

third terms on the right-hand side of the equation (33) indicate (surprisingly) that the ΔG^0_{ads} value for a given substance is substantially independent of the metal.[125]

(ii) Dependence of the Equilibrium Adsorption on the Inhibitor

On the other hand, U in equation (33) becomes more negative as the solubility of the organic substance in water decreases, i.e., the adsorbability increases with decreasing

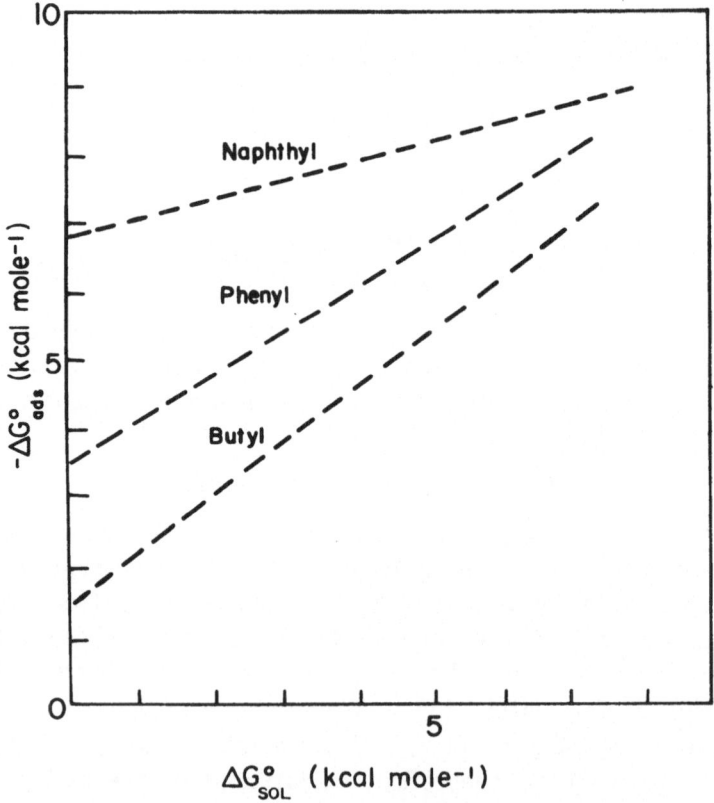

Fig. 38. Experimental relationship (schematic) between standard free energies of adsorption and solution for naphthyl, phenyl, and butyl derivatives (Blomgren, Bockris, and Jesch[112]).

solubility. In fact,[112,125]

$$U = U' - \Delta G_{\text{sol}}^{0} = U'' + RT \ln \frac{c_s}{p_0} \qquad (34)$$

where p_0 is the vapor pressure of the pure organic substance, c_s is the concentration of organic substance in the saturated solution (i.e., the solubility), and $\Delta G_{\text{sol}}^{0}$ is the standard free energy of solution of the organic substance in water from the vapor phase. Thus, equations (33) and (34) show that a linear relation exists between $-\Delta G_{\text{ads}}^{0}$ and $\Delta G_{\text{sol}}^{0}$, and this has been found to hold for the adsorption of a series of organic substances on Hg^{112} (see Fig. 38).

(iii) The Dependence of the Coverage of the Inhibitor upon Current Density

However, when metal deposition occurs at the electrode surface, the amount of substance adsorbed may be determined by nonthermodynamic factors. If it is assumed that the adsorbed particles are incorporated into the growing crystal and that the freshly deposited metal exposes a perfectly clean surface to the solution, the maximum surface concentration of impurities can be calculated from the average lifetime of the recently formed surface before it is covered by the next layer of metal. This lifetime is simply given by

$$\tau_{\text{surface}} = \frac{Fh}{Vi} \qquad (35)$$

where F is the faraday, i is the current density, V is the molar volume of the metal, and h is the height of an atomic layer. Taking $V \sim 10\ \text{cm}^3/\text{mole}$, $h \sim 3 \times 10^{-8}\ \text{cm}$, and $i = 30$ mA/cm^2, τ_{surface} will be about 10^{-2} sec.

With respect to the kinetics of the adsorption process, two steps must be considered: a mass transport step, i.e., transfer of the adsorbable species from the bulk of the solution to the layer in contact with the electrode surface; and an adsorption step, i.e., transfer of the particle from the solution in the pre-electrode layer to a point on the surface where it remains adsorbed; it will be assumed that the second step is very fast and, hence, the first step mentioned is rate determining. Two principal cases must be considered when analyzing the rate of mass transport to the electrode surface:

1. The case in which a pure diffusional process takes place without the interference of convective transport—a situation which will prevail at sufficiently short times in regard to the onset of natural convection.

2. The case in which a mixed diffusion and convection mechanism is responsible for the supply of particles to the electrode—the predominant situation at longer times or in the presence of mechanical stirring.

For case 1 the time dependence of the amount adsorbed is given by[145,146]

$$\Gamma(t) = 2c_b \left[\frac{Dt}{\pi} \right]^{1/2} \tag{36}$$

if the adsorption situation is far from equilibrium, i.e., if $\theta(t)/[1 - \theta(t)]^n \ll \theta_{eq}/(1 - \theta_{eq})^n$. This can be exemplified by the following typical values: $D = 10^{-5}$ cm^2/sec, $\Gamma_{max} = 10^{-9}$ mole/cm^2, $k = 10^8$ cm^3/mole (corresponding to $\Delta G^0_{ads} \approx -4$ kcal/mole), $c_b = 10^{-7}$ mole/cm^3, and $t = \tau_{surface} = 10^{-2}$ sec. Equations (31) and (36) give $\theta_{eq} \approx 0.9$ and $\theta(\tau_{surface}) \approx 0.04$.* Hence, the situation is quite far from equilibrium and, as shown by equations (35) and (36), $\Gamma(\tau_{surface})$ is totally independent of ΔG^0_{ads} and proportional to $i^{-1/2}$.

If the situation involves convective transport (and this is more likely to be the case when there is simultaneous deposition of a metal on the electrode), the equation governing the rate of adsorption is[146]

$$\Gamma(t) = \Gamma_{eq} \left[1 - \exp\left(-\frac{Dt}{\delta K} \right) \right] \tag{37}$$

where $K = k\Gamma_{max}$. Equation (37) holds for a linear approximation of the adsorption isotherm, i.e., if $\theta \approx kc_e$ (c_e is the concentration of adsorbable substance in a layer of solution adjacent

---

* This is the maximum coverage reached during the lifetime of a freshly deposited surface at a current density of 30 mA/cm^2.

to the electrode surface). For the same value of the parameters assumed earlier ($D = 10^{-5}\,\text{cm}^2/\text{sec}$, $K = 10^{-1}\,\text{cm}$, $t = \tau_{\text{surface}} = 10^{-2}\,\text{sec}$), and taking $\delta = 10^{-2}\,\text{cm}$,

$$\frac{D\tau_{\text{surface}}}{\delta K} = 10^{-4} \ll 1$$

Hence, a linear approximation of (37) is valid:

$$\Gamma(\tau_{\text{surface}}) \approx \Gamma_{\text{eq}} \frac{D\tau_{\text{surface}}}{\delta K} \tag{38}$$

But

$$\Gamma_{\text{eq}} = Kc_b$$

Hence

$$\Gamma(\tau_{\text{surface}}) = \frac{c_b D\tau_{\text{surface}}}{\delta} \tag{39}$$

which again shows $\Gamma(\tau_{\text{surface}})$ to be independent of ΔG^0_{ads} but proportional to i^{-1} [because of equation (35)]. The physical meaning of the independence from ΔG^0_{ads} is that, when the adsorption process is far from equilibrium and the rate of adsorption is controlled by the mass transport process, every particle arriving at the electrode surface is adsorbed immediately, and the probability of its desorption is negligible. The boundary condition at the electrode surface is, thus, independent of the thermodynamic parameters. As the concentration of the organic substance in the solution increases (or the current density decreases), τ_{surface} becomes a too long time for (36) or (38) to be good approximations of the kinetics of adsorption because thermodynamic equilibrium is approached. Therefore, ΔG^0_{ads} becomes important at higher concentrations of the adsorbable species (or lower current densities), i.e., adsorption equilibrium is more nearly approached during the lifetime of the surface.

The use of equations (36) and (37) is open to objections. Equation (36) is based on the initial condition $c(x, 0) = c_b$

which ignores the fact that diffusion of the adsorbate to a freshly deposited, clean, metal surface occurs in the already existing diffusional field connected with the transport of the inhibitor to the previously exposed metal surface. In other words, the creation of a clean metal surface at $t = \tau_{\text{surface}}$ does not imply a return to the conditions at $t = 0$ in respect to the concentration profile [which is the assumption underlying the use of equation (32)]. On the other hand, equation (37) is based on the assumption of adsorption equilibrium at the metal surface, $\Gamma(t) = Kc(0, t)$. Since $\Gamma(t)$ is not uniform in the direction of propagation of growth layers on the surface, $c(0, t)$ will also be nonuniform, and, therefore, the model of unidimensional diffusion breaks down. This objection disappears when the linear approximation of the exponential is adopted [equations (38) and (39)]; in fact the linear approximation is equivalent to a negligible probability of desorption of adsorbed particles [$c(0, t) = 0$], i.e., a "far from equilibrium" situation.

4. MECHANISMS OF INHIBITION

There is no single mechanism of inhibition because the term covers, in the present context, the totality of effects which arise in electrocrystallization as a result of entities adsorbed upon the surface (apart, of course, from the blocking effects caused by the adions during transfer to a growth site). Thus, inhibition (as has often been stressed by H. Fischer[55]) is (almost) an *essential* part of the electrocrystallization process, in that, as it seems at present, most explanations of electrocrystallization processes involve the assumption that the elementary processes are affected by adsorbed entities, usually adventitious ones. In this "pan inhibitor" view point, even the adsorbed water molecules, and perhaps the adsorbed anions, are "inhibitors" in the sense that, as adsorbed materials, they effect the growth of crystals on electrodes.

Consequently, it seems convenient to analyze the effect of adsorbed substances upon the different steps of the electrocrystallization process. The following actions may occur as a result of the presence of adsorbed material:

1. The free energy of activation for the elementary charge transfer steps can be modified. Hence, the rate constant i_0, the rate determining step, and even the path of the reaction will depend upon the amount and nature of the adsorbed substance. Superimposed upon this, there is the purely geometrical effect of reduction in the area available for charge transfer to $(1 - \theta)$ of the total area.*

2. The mean free path for lateral diffusion of the adions will be shortened by the presence of other adsorbed particles. This is equivalent to a decrease in the diffusion coefficient of adions which can result in: (a) an increase in surface diffusion control; (b) an increase in adion concentration at steady state to a point such that the rate of two-dimensional nucleation becomes appreciable and, hence, a *decrease* in surface diffusion control due to the reduction in the distance between growth steps.

3. If there is preferential adsorption at growth steps,† the surface concentration at these points can considerably exceed the value calculated in terms of the total (geometric) area. Perhaps this would explain the extremely high sensitivity of the growth form with respect to very small amounts of added impurities (while the electrochemical parameters, usually reflecting the energetics of charge transfer, exhibit much less sensitivity). It is difficult to predict the effect of the blocking of growth sites. Several possibilities will be listed: (a) a decrease in the edge free energy and hence in the radius of the critical nucleus [cf. equation (7) p. 61]; this is equivalent to an increase in the "activity" of growth sites, reduction of the "edge overpotential," and decrease in the distance between growth steps (hence, decrease in the surface diffusion overpotential); (b) quite oppositely, the adsorption of particles may completely block growth sites, thus increasing the distance

---

* This is the case if we assume that the same sites are involved in adsorption and in charge transfer.

† cf. Discussion on p. 116.

between steps, or it may force the growth steps to present regions of large curvature, thus producing an "edge overpotential," increasing the adion concentration at the steps and facilitating nucleation at the midplanes.

4. If the distribution of the adsorbed substance is heterogeneous and the coverage large enough to produce high local current densities (in the regions of the surface free from adsorbate), an appreciable concentration overpotential may be generated. Even when a uniform distribution of adsorbate exists under equilibrium conditions, there are reasons to believe that a nonuniform coverage arises when deposition is taking place. In fact, if the adsorbate is being incorporated into the growing crystal, a steady state both in growth and coverage may well involve regions of high activity (with respect to deposition) and low inhibitor coverage, and regions of low activity and high inhibitor coverage, i.e., a "kinetic heterogeneity" would exist.*

The preceding discussion has a qualitative and rather speculative character. The ideas involved require further elaboration; more experimental data are needed, especially regarding the coverage *during* deposition, since, as pointed out earlier, this might be quite different from the equilibrium value. With respect to the effect of the additives upon the morphological properties of the deposit, several aspects have been already mentioned. (a) Grain refinement can be attributed to the increased frequency of nucleation brought about by the inhibition of, e.g., surface diffusion (resulting in a higher adion concentration). Vermilyea[147] offers the following explanation for grain refinement: Initially, the surface is covered by impurities which block the growth steps; when the current is switched on, nucleation soon occurs, and deposition takes place at a high rate on the newly formed nuclei; this high local current density cannot be sustained for a long time, due to depletion of cations and increase in the diffusion layer

---

* This kind of heterogeneity would be responsible for the formation of whiskers in the presence of impurities, according to the model of Price, Vermilyea, and Webb (*cf.* Chapter 8, Section 3).

thickness as the crystallite grows; as the current density decreases locally, a considerable degree of coverage by an inhibitor takes place. Thus, the poisoning of the (initially clean) surface of the crystallite prevents further growth, and nucleation will occur again at a point on the surface far from the previously growing nucleus. (*b*) The bunching of micro-steps and formation of macroscopic (or even visible) layers have been linked to the (kinetically controlled) adsorption of impurities (*cf*. Chapter 10). (*c*) Two important technological applications of the effect of the additives, leveling and brightening action, merit special reference and they will be treated in the next chapter.

Chapter 12

The Theory of the Electrochemical Leveling and Brightening of Surfaces

1. FACTORS AFFECTING CURRENT DISTRIBUTION

The appearance of an electroplated surface depends largely upon its uniformity. Hence the interest in the study of current distribution at an electrode of irregular profile, e.g., on a coarsly polished metal surface.

It is possible to distinguish three main sources of uneven current distribution:

1. Structural, that is, related to the variation in kinetic parameters with crystal face.[113,127] Different parts of the electrode will usually exhibit differently oriented surfaces; the path, rate determining step, and rate constant will, in general, change from one facet to another, and the resulting uneven distribution of current gives rise to the process known as faceting.

2. Diffusional, that is, originating from the different rates of mass transport from the bulk of the solution to different points on the electrode surface. Although the exact mathematical treatment requires the solution of Fick's second law with the appropriate boundary conditions, the physical model can be conveniently expressed in terms of the variations in the effective thickness of the diffusion layer δ_N over the electrode surface: δ_N is larger at recesses and smaller at peaks and, hence, the current density at peaks is larger than

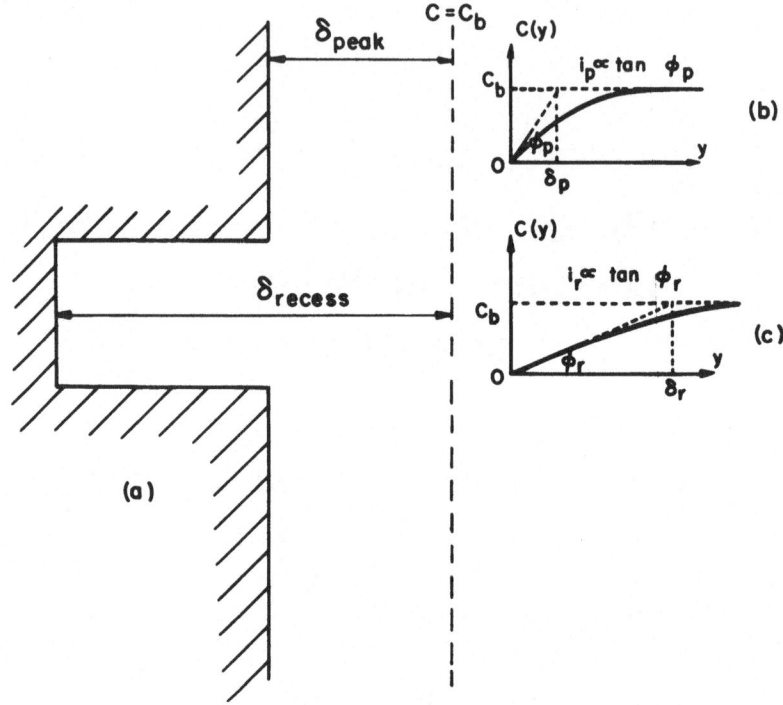

Fig. 39. Schematic representation of the causes of uneven current
distribution at an irregular electrode (a) under limiting-current
conditions. Concentration profiles at a peak and a recess are
given in (b) and (c), respectively.

that at recesses, but only if the mass transport process exerts
any control on the rate of the deposition (Fig. 39). Thus,
maximum nonuniformity tends to occur under limiting current
conditions ($c_e = 0$ or, more precisely, $c_e \ll c_b$ and then $c_b - c_e \approx c_b$
at all positions near the electrode surface). As the situation
departs from diffusion control, the nonuniformity of the
deposition decreases (Fig. 40).

The preceding discussion rests upon the assumption of
flatness of the outer boundary of the diffusion layer, i.e., the
inability of the hydrodynamic flow to penetrate the irregularities
of the surface. The validity of the case considered is thus
limited to profiles exhibiting irregularities smaller than the

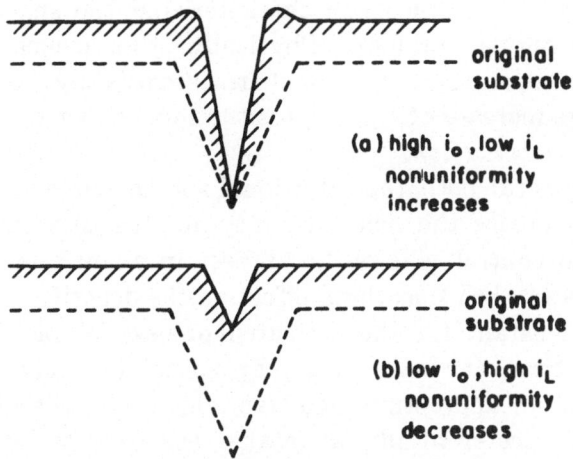

original
substrate

(a) high i_o, low i_L
nonuniformity
increases

original
substrate

(b) low i_o, high i_L
nonuniformity
decreases

Fig. 40. Schematic representation of deposition with (a) bad and (b) good microthrowing power.

average thickness of the diffusion layer, that is, limited to *microprofiles*. As the absolute dimensions of the original irregularities increase, the boundary layer follows the profile, i.e., no further increase of irregularity occurs, even under diffusion control.* In fact (see p. 130), some smoothing of the surface will occur under these circumstances.

3. Ohmic resistance in the solution as a cause of non-uniform current distribution. If there is an appreciable difference in the solution resistance from the bulk of the solution to different points on the electrode surface, the distribution of current density will be such as to favor those sites which offer the path of least resistance; that is, peaks will receive a larger amount of depositing material than recesses, and hence the roughness of the surface will progressively increase. This situation is known as *primary current distribution*.[161,165] The tendency will be counteracted if the differential reaction resistance $|d\eta/di|$ is large compared with a/κ, which is a measure of the *difference* in solution resistance from the bulk to peaks and to recesses (κ being the conductivity of the

* This will be the case unless the effect of the irregular distribution of electrical field becomes preponderant.

solution and a a dimension characterizing the shape and size of the irregular profile). Physically, this means that any trend to make $i_{peak} > i_{recess}$ will be immediately compensated by a large increase of $| \eta_{recess} |$, a situation known as *secondary current distribution*.[161]

In general both the diffusional and the ohmic control will tend to increase the nonuniformity of the substrate, while activation control will result at best in a uniform deposition without a marked smoothing effect on the deposit. Only when the rate constant for the activated process is made larger at recesses than at peaks (e.g., by adsorption under diffusion control of an appropriate additive which will, therefore, tend to adsorb preferentially at peaks—see next section), will a rapid "leveling" of the surface take place.

In summary, an irregular current distribution arises from diffusion controlled deposition at small irregularities (micro-profiles); the uniformity of crystal growth improves as the dimensions of the profile increase and the diffusion layer "penetrates" the crevices in the surface. On larger irregularities (macroprofiles, in which the diffusion layer "follows" the surface irregularities) ohmic control of the current distribution results in nonuniform deposits. In these conditions, any factor which would cause an increase in the polarization resistance (including the onset of some diffusional control) will improve the uniformity by effecting a changeover from primary to secondary current distribution.[149,166,167]

2. LEVELING

The phenomenon of leveling is the progressive reduction of the irregularities of an electrode surface upon deposition. It is necessary to distinguish between two kinds of leveling processes: (a) geometric leveling (Fig. 41) which accompanies a uniform current distribution and is due to the convergence of material at recesses and divergence at peaks, resulting in a gradual "filling up" of the former points; (b) "true" or "electrochemical" leveling arising from a nonuniform current

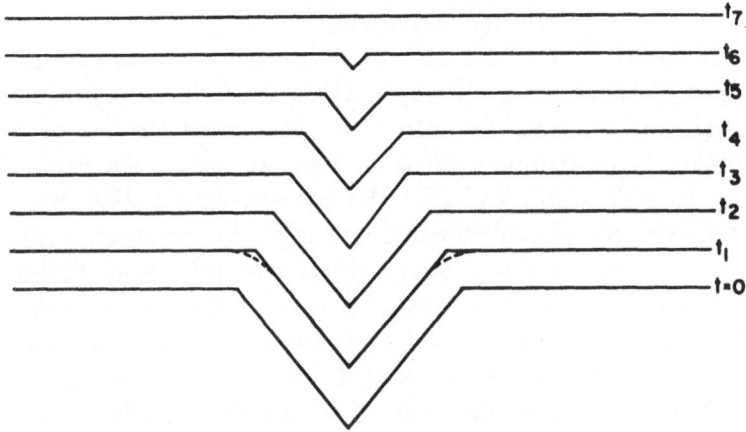

Fig. 41.   Geometric leveling with gradual elimination of irregularities
when uniform deposition takes place.

distribution such that $i_{\text{recess}}/i_{\text{peak}} > 1$. Obviously, a much faster
leveling action will take place in the second case than when
the current distribution is uniform.

Since all the mechanisms mentioned in Section 1 of this
chapter predict either a uniform distribution (i.e., a slow
decrease in surface irregularities), or a larger rate of deposition at peaks, and also because of the experimental
correlation between leveling and the effect of addition agents
upon current–potential curves, a mechanism in which a primary
role is assigned to the addition agent must be invoked in the
explanation of the leveling action. Kardos[167] proposed a
mechanism based on the following assumptions:

1.   The rate of deposition is activation-controlled over
the whole surface.

2.   At a given current density, an increase in coverage
by the leveling agent results in a strong *increase* in the over-
potential, that is, $d|\eta|/d\theta$ is large and positive (in other words,
the current density *decreases* with increasing coverage at
constant potential).

3.   The surface coverage by the additive is not the

equilibrium one but is controlled by the rate of diffusion of the additive toward the electrode.*

According to 3, the surface coverage at recesses will be smaller than at peaks if the surface irregularities are small enough not to have an effect on the outer profile of the diffusion layer (thinner diffusion layer at peaks, see Fig. 39). Then from 2 it is clear that the current distribution will be as required for electrochemical leveling. Considerable support for the preceding model arises from the following findings: (a) The amount of incorporated leveling agent (as revealed by radiotracer techniques[173] or by chemical etching[151]) gradually decreases from the micropeaks toward the microrecess. (b) The current density obtained at constant electrode potential *on a flat metal surface* decreases with increasing concentration of leveling agent and with increasing agitation rate[150,151,168,174] (at the rotating disk electrode[150,174] with increasing $c_b \times \omega^{1/2}$, where c_b is the bulk concentration of the leveling agent and ω is the angular velocity of the disk). (c) The rate of consumption of the leveling agent shows the features of diffusion control[150,168,175] and is, for coumarin in nickel deposition at the rotating disk electrode, proportional to $c_b \times \omega^{1/2}$. (d) With pulsed current, leveling disappears if the duration of the cathodic impulse becomes too short (at constant and sufficiently large ratio of interruption-time/impulse-time).[174] (e) The model explains all the features of leveling, including the occurrence of leveling maxima with increasing c_b or $(c_b \times \omega^{1/2})$.[150,151,173,176]

3. BRIGHT ELECTROCRYSTALLIZATION

In any attempt to analyze the mechanism responsible for the development of bright electrodeposits, it is a prerequisite to understand the structural factors which determine the degree of brightness of a surface.

---

* This also implies the assumption of "irreversible adsorption", i.e., as deposition proceeds, the adsorbed substance is consumed at the electrode, either by burying or by electrochemical reduction to some product which desorbs or has no effect on the electrodeposition process.

A historical difficulty in dealing with this problem is the lack of a clear definition of the word "brightness" in terms of physical magnitudes. Thus, most of the early work has centered around a vague definition in terms of visual appearance. More recently, the reflecting power, as measured by the amount of light specularly reflected off a surface,* i.e., at an angle equal and opposite to that of incidence with respect to the normal to the geometrical surface, was used to express the brightness of a surface.[152]

A more precise definition, not involving the actual reflectivity of the surface, would be in terms of the ratio between specularly and diffusely reflected light.[153]

The early attempts to establish a correlation between brightness and structure of a surface met with little success. Often, contradictory observations were reported. This can be attributed to the fact that the conditions of electrodeposition which produce a surface with the required structural characteristics for a bright appearance, also result in other (non-essential) morphological characteristics, which are then assumed to be responsible for the brightness, while in fact they are only incidentally connected with it.

Thus, it has been found that a bright electrodeposit is often related to a grain refinement[154]; however, Clark and Simonsen[155] reported no correlation between grain size and brightness, and the results of Leidheiser and Gwathmey[156] for Ni electrodeposition on single crystal spheres show a mirror bright single crystal deposit (on the {100} face) and a dull polycrystalline growth (on the {111} face). Hence, *there is no fundamental relation between grain size and brightness.*

The experimental evidence concerning the dependence of the brightness of an electrodeposit upon the presence of any (or a particular) preferred orientation clearly indicates the absence of such dependence,[152,154] although the electroplating conditions leading to a bright deposit usually prevent the development of a preferred orientation. This point will be

* Or, more precisely, the illumination of the monitoring detector at constant illumination of the reflecting surface.

further discussed in relation to the mechanism of formation
of a bright deposit.

If an attempt is made to define brightness of a surface
as the ability to reflect light *only* in a direction making an
angle with the normal equal to that of the incident beam (i.e.,
approach to "ideal mirror"), then the surface will be increasingly
bright as the scattering introduced in the reflected light by
the surface irregularities decreases, that is, as the deviations
of the actual surface from the ideal, geometrical boundary
are eliminated. The question arises as to what scale of surface
roughness is responsible for the introduction of scattering in
the reflected light; the answer can be given on intuitive
grounds. A surface will be effectively smooth from the point
of view of reflectivity when the order of magnitude of the
irregularities becomes less than the wavelength of the radi-
ation being used.[182] Thus, Weil and Paquin[152] found a linear
relationship between the logarithm of the amount of light
specularly reflected by a surface (under well-standardized
conditions) and the fraction of the surface presenting a roughness of less than 1500 Å, and electromicrographs of electro-
polished surfaces show no large irregularity or faceting.

It is therefore apparent that in order to account for the
formation of bright electrodeposits it is necessary to discuss
the mechanistic aspects concerning the roughening of a growing
surface.

It was mentioned in connection with the phenomenon of
texture (Chapter 9) that the different rates of growth of
crystallographic faces is the determining factor leading to
faceting of a growing surface and, hence, to its roughening.
It is natural, then, to attempt to explain the action of brighteners in terms of an equalization of the rates of growth of
the different crystal planes present on the surface. There
are many experimental facts which indicate that the diffusion-
control leveling mechanism described in Section 2 does not
operate at a crystallographic level (in particular, the lack of
leveling properties of many organic additives that produce
bright deposits, as well as the observation that the concen-

tration of a given organic substance required to produce leveling differs greatly from that required for brightening action).[149] Hoar has proposed two possible mechanisms whereby equalization of the growth rate on different crystal planes and, consequently, brightening would occur.[157]

1. Assume that the organic substance (brightener) is reversibly adsorbed at the surface forming a nearly complete monolayer; the kinetic character of the adsorption equilibrium would result in the continuous formation and destruction of holes in the adsorbed layer, where deposition of the metal ions can take place. Since such holes are randomly distributed, a completely uniform (at a crystallographic level) deposition would take place. Hence, no faceting of the surface would occur, and any facets initially present would be gradually eliminated according to a geometric leveling mechanism.

2. An alternative explanation is based on the heterogeneity of the surface with respect to adsorption. If the more active faces with respect to deposition also constitute preferred sites for adsorption, and if the adsorbed substance is an inhibitor, i.e., it interferes with the deposition process, then the inhibition will be higher at the more active faces, and the required equalization of the rates of growth at different points on the surface is attained. Again absence of faceting and elimination of microroughness through geometric leveling would lead to bright deposits.

The assumption that brightening of a surface occurs when the growth is uniform at a crystallographic level is consistent with some observations concerning the relationship between brightening and development of texture; clearly, the condition of uniform growth is satisfied either when faceting is absent and, hence, no preferred orientation develops[56] (see Chapter 9), or when only one crystal face is present on the surface, i.e., when a smooth single crystal surface grows or when a very strong preferred orientation is developed. Correspondingly, bright electrodeposits were observed both in the case of deposits with a strong texture (see Ref. 153) and in the absence of any preferred orientation,[154] as well as for smooth single crystal faces.[156] It is interesting to note that, on a random polycrystalline, polished substrate, the deposition time at which dimming of the light reflected by the surface is first observed, is approximately the same—and varies with the current density in the same way—as the time at which a preferred orientation develops,[158] i.e., faceting and predominance of certain faces is attained.

The decrease in grain size often observed in the formation of bright deposits can be explained in terms of the increased probability of nucleation under conditions of high inhibition; however, a higher nucleation rate would not constitute the cause of a brighter deposit, being only a parallel phenomenon. Experimental results obtained by Ke *et al.*[159] for adsorption of thiourea and for Cu deposition (with and without thiourea added) on different faces of a Cu single crystal sphere lend support to the hypothesis of a uniformly adsorbed layer and equalization of the growth mechanism at different points of the surface.

It is clear that knowledge of the dependence of the adsorption of organic brighteners upon potential and its effect on the reaction mechanism and rate, together with data concerning the relations between surface heterogeneity and adsorption, is necessary before the mechanism of brightening can be successfully approached.

Chapter 13

Electrocrystallization and Crystallization from the Gas Phase

The problem of crystal growth from the vapor has been analyzed in detail, especially by Frank and his collaborators.[12,44,178,179] In particular, they considered the problem of the equilibrium structure of a crystal surface and concluded that the steps on the surface of a crystal would present a large concentration of kinks and, therefore, would readily incorporate adatoms in their near vicinity. Another important conclusion concerns the primary role of surface diffusion in the mechanism of evaporation and growth of crystals.

Frank[179] was the first person to present the idea that the growth of crystals must be associated with the presence of imperfections on the crystal surface, in particular, screw dislocations (see Chapters 2 and 10). These would give rise to self-perprtuating steps on the crystal surface, thus obviating the necessity of assuming two-or three-dimensional nucleation as a necessary step in the growth process. Burton, Cabrera, and Frank[44] considered the growth forms which would arise from a single screw dislocation, from pairs of dislocations, and from aggregates of dislocations.

Vermilyea[45] was the first to elaborate this type of model with respect to the electrolytic growth of crystals. He derived i-η equations for deposition under steady-state con-

Table 9

Parameters Controlling Crystal Growth from the Vapor and Electrolytic Growth

Parameter	From the vapor[44]	Electrolytic[13]
Supersaturation ratio in the bulk phase (σ)	p/p_0	$\exp\left[-\dfrac{zF\eta}{RT}\right]$
Supersaturation ratio on the surface	c_{ad}/c_0	c_{ad}/c_0
Mean lifetime of adsorbed particles	$\nu^{-1}\exp[W_s/RT]$	$\dfrac{zFc_0}{i_0}\exp\left[\dfrac{(1-\alpha)zF\eta}{RT}\right]$

ditions and with growth sites arising from isolated screw dislocations.

Although the basic model for electrolytic crystal growth is considered to be the same as that for crystal growth from the vapor—deposition on flat surfaces followed by surface diffusion to, and incorporation at, growth sites—it is necessary to remember some important differences between both processes,

1. The driving force in the case of crystal growth from the vapor is simply given by the supersaturation, i.e., a concentration of particles in the gas phase larger than that corresponding to equilibrium. In the case of electrolytic growth an externally controllable parameter is introduced: The applied overpotential provides the required driving force (see Table 9).

2. The rate of incorporation of particles at the crystal is slower in the electrocrystallization case than in that of growth from the gas phase. For example, utilizing Hertz equation[160] for the rate of bombardment on a plane surface, the rate of deposition is about 10^{-3} mole/sec cm², * whereas in the electrochemical case, it is about 10^{-5} mole/sec cm² even at 1 A/cm², an unusually high current density.† In spite of this, the mass-transport process is more likely to be rate-determining

___

* Assuming a molecular weight of 100, a vapor pressure of 0.04 atm, a condensation coefficient of 0.1, and a supersaturation ratio of 1.1.

† From Table 9, the overpotential corresponding to $\sigma = 1.1$ is about -2.5 mV; hence, to account for a current density of 1 A/cm², i_0 should be 10 A/cm² (from $\eta = -iRT/i_0F$).

in the electrochemical case because of the higher diffusion coefficient in the gas phase (10^{-1}–1 cm^2/sec) compared with that in solution ($\sim 10^{-5}$ cm^2/sec).

3. The particles involved in the electrodeposition are charged and hydrated and one (or several) charge transfer and dehydration steps must occur, i.e., the energetics of the interfacial transfer depends significantly upon the applied potential difference. Rate-determining passage across the double layer is possible for the electrochemical case. The field strength in the double layer is about 10^7 V/cm and, therefore, some electrostrictive distortion is probable.[103]

4. Surface diffusion is likely to be slower in electrocrystallization than in the deposition from the vapor phase because of adsorption on the electrode of water molecules and ions.[55,103]

5. In the case of electrocrystallization, the particles undergoing surface diffusion are hydrated as a result of their partial ionic character (see Chapter 4, Section 7). This will affect considerably the rate of surface diffusion; the heat of activation will be lower and the entropy of activation higher than in the absence of the hydration sheath (vapor case); the stabilization of the adsorbed state (as a result of the hydration energy) will also result in a larger equilibrium concentration of adions.

These differences will affect both the metal deposition and crystal growth aspects of the subject. The general tendency is for the electrocrystallization case to be slowed down and affected by more factors, at the same degree of supersaturation, and hence to be a somewhat more complicated phenomenon than that of crystal growth from the vapor.

Chapter 14

Prospects

In order to judge the prospects of the field of electro-crystallization, it is best to begin with some classification of research in this area. Thus, A. K. Reddy has suggested the following divisions:

1. Basic factors, 1–10 Å (atomic level): The work referred to in this monograph as "metal deposition."[51]

2. 10–1000 Å (crystallographic level): The work referred to here as "crystal growth." It takes into account the growth velocities of different crystal faces and the effect of adsorption at growth sites of adventitious impurities (e.g., work on inhibitor effects).[51,55]

3. 0.1–100 μ (diffusion-layer level): The dependence of growth forms upon local diffusion gradients (e.g., micro-throwing power, theory of leveling).[149]

4. Laplace-equation level: This research concerns macro-current distribution and, therefore, shape of the deposit, depending on the influence of the local solution resistances and the fraction of the total overpotential which they make up (e.g., theory of throwing power).[161]

Out of the advances in these fields, particularly the first three, many technological advances can be anticipated. For example, it should be possible to make still further progress in the field of high-speed electrodeposition by application of our increasing knowledge of the hydrodynamics of the solid–solution boundaries. It should also be possible to obtain a maximum effect of additives on the metal surface with respect to brightening and leveling, with a minimum of embrittling codeposition. Correspondingly, our knowledge of the rates of deposition at growth sites, when wedded to our small quantitative knowledge of the kinetics of crystal growth, will provide the possibility of making large single crystals and, possibly, controling and greatly diminishing the defect concentration in the final product. The production of metals of a strength much higher than those hitherto produced will thus be possible. Also, in the field of electrodeposition from nonaqueous solutions (including molten salts) there has been empirical efforts for many years without the benefit of any theory of the solutions involved.

A few of the research goals of the next few years may be mentioned:

1. The greatest need is in methods for watching crystal growth on a surface. At present, Nomarski optics allows the observation of growth approximately 100 Å high and 2500 Å in length. Essentially, this means that macrosteps can be examined. However, it does not mean that it is possible to examine the microgrowths which occur between the steps[80] and account for much of the current. For this, an electron-microscopic type of approach is desirable. The difficulty, of course, is the absorption of electrons by the solution, i.e., an electron microscope of very greatly increased accelerating power would be needed. However, such a microscope has already been built in France. The availability of a tool of this kind would greatly change the field.

2. Extensive use of transients, particularly potentiostatic, for an examination of both the metal deposition and crystal growth processes.[69] The simultaneous use of such methods with measurements of the impedance of the interface at various frequencies is desirable. Such measurements, made with variation of concentration of the reacting entities in solution, make possible, in principle, the separate evaluation of rate constants for metal deposition and crystal growth, from which detailed mechanistic conclusions should result.

3. The role of impurities is as yet largely hypothetical.

They are invoked in practically all mechanisms in metal deposition, but there have been only beginning attempts to examine such effects rationally, i.e., to increase the amount of *adsorbed* impurities and correlate this with the observed characteristics of deposition process.[80,133] In particular, at present there is no explanation of the presence of *visible* steps except in terms of impurity adsorption, and there is little quantitative experimental work to confirm or deny that hypothesis.

4. A very general result is the "passivation" of freshly deposited surfaces after a time of a few seconds. The conclusion reached concerning its cause[68] (migration of adions from one crystal plane to another, i.e., "settling down" of the surface) should be examined both in respect to single crystals and controlled purity addition.

5. Electrode kinetic theory indicates[39] that mechanism analysis is greatly facilitated if the parameters of a reaction are known for *both directions*. This has very seldom been taken into account by workers in the field.[61,67,181]

6. Parallel studies of the complex ions formed in solution, and their equilibria with simple entities, as also the adsorption of all entities of the solution on the electrode, appear to be essential prerequisites in the understanding of electrocrystallization. For example, the adsorption of inorganic cations on solid metals is unexpectedly large and relatively independent of potential.[162]

7. The use of whiskers, deposited from the vapor phase (pioneered by Vermilyea),[58] has considerable potential as a research tool, because it would allow the easy variation of the number of dislocations on the surface.

8. The theories of dendritic and whisker growth should be further examined. Several theoretical points remain unsolved and the crystallographic aspects are not yet understood. The study of these growth forms would greatly help in the understanding of the basic mechanisms of electrocrystallization.

9. The mechanics of step movements, as a function of concentration of solute ions and additives, must be examined.

This is greatly facilitated by the use of Nomarski optics.

10. Correspondingly, at least 10% of the research force in electrocrystallization must be used in continuing purely theoretical studies.

Though the generalities concerning the monolayer metal deposition stage are well understood, and a fairly extended knowledge of the phenomenological aspects of crystal growth exists, a number of problems remain unanswered, and the mechanistic link between monolayer deposition and crystal growth still has to be found.

In connection with epitaxial growth on single crystals, the following basic questions arise:

11. How do the path, rate determining step, and kinetic parameters depend on the structure and recent history of the metal surface, e.g., on the particular crystal face exposed at the interphase?

12. What is, at any stage during the growth process, the distribution of crystal faces at the growing surface, and how does this distribution affect the deposition?

13. What is the current distribution at a surface composed of different facets?

14. How does a given current distribution affect the growth and, hence, the further distribution of crystal faces at the interphase?

15. How should the local effects (e.g., current distribution) be integrated to obtain the macroscopic (measurable) electrochemical parameters?

The answers to these questions, both theoretical and experimental, will provide the solid base upon which a theory of electrolytic crystal growth can be built.

References

1. Faraday, *Phil. Trans. Roy. Soc. London* **124** (1834) 77.
2. Gibbs, *Collected Works*, Vol. I, footnote p. 325, Longman's, New York, 1928.
3. LeBlanc, *The Elements of Electrochemistry*, p. 250, Macmillan, London, 1896.
4. Huntington, *Trans. Faraday Soc.* **1** (1905) 324.
5. Glocker and Kaupp, *Z. Physik* **24** (1924) 121.
6. Bockris and Conway, *Record Chem. Progr.* **25** (1964) 31.
7. Thompson, *Proc. Roy. Soc. London, Ser. A* **133** (1931) 1.
8. For a summary of the work of this school, see: Finch, Wilman, and Yang, *Discussions Faraday Soc.* **1** (1947) 144.
9. Fischer, *Elektrolytische Abscheidung und Elektrokristallisation von Metallen*, Springer-Verlag, Berlin, 1954.
10. Taylor, *Proc. Roy. Soc. London, Ser. A* **145** (1934) 362.
11. Burgers, *Proc. Acad. Sci. Amst.* **42** (1939) 293.
12. Burton, Cabrera, and Frank, *Nature* **163** (1949) 398.
13. Mehl and Bockris, *Can. J. Chem.* **37** (1959) 190.
14. Erdey-Gruz and Volmer, *Z. Physik. Chem. Leipzig A* **157** (1931) 165.
15. See, for example, Azaroff, *Introduction to Solids*, (particularly Chapter VII), McGraw-Hill Book Co., New York, 1960.
16. Erdey-Gruz and Volmer, *Z. Physik. Chem. Leipzig A* **150** (1930) 203; *cf.* Bockris, *History and Problems of Electrode Kinetics*, Proceedings of the First Australian Conference on Electrode Kinetics, Pergamon Press, New York, 1964.
17. Bockris and Conway, *Trans. Faraday Soc.* **45** (1949) 989.
18. Damjanovic, Paunovic, and Bockris, *Electrochim. Acta* **10** (1965) 111.
19. Frank, *Growth and Perfection of Crystals*, Edited by Doremus, Roberts, and Turnbull, p. 411, John Wiley & Sons, New York, 1958.
20. Cabrera and Vermilyea, *Growth and Perfection of Crystals*, Edited by Doremus, Roberts, and Turnbull, p. 393, John Wiley & Sons, New York, 1958.
21. Pentland, Bockris and Sheldon, *J. Electrochem. Soc.* **104** (1957) 182.
22. Barker, *Transactions of the Symposium on Electrode Processes*, Edited by Yeager, p. 366, John Wiley & Sons, New York, 1961.

[23] Bockris, *Modern Aspects of Electrochemistry*, Vol. 1, Edited by Bockris, Chapter 4, Butterworths, London, 1954.

[24] Volmer, *Physikal. Z. Sowjet.* **4** (1933) 346.

[25] Rojter, Juza, and Polujan, *Acta Physicochim. URSS* **10** (1939) 389.

[26] Bowden and Rideal, *Proc. Roy. Soc. London, Ser. A* **120** (1928) 59.

[27] Baars, *Sitzb. Ges. Beförd. Ges. Naturw.* **63** (1928) 213.

[28] Aten and Boerlage, *Rec. Trav. Chim.* **39** (1920) 720.

[29] Kossel, *Nachr. Ges. Wiss. Göttingen* **1927**, 135.

[30] Stranski, *Z. Physik. Chem. Leipzig* **136** (1928) 259.

[31] Volmer, *Das Elektrolytische Kristallwachstum*, Herman et Cie, Paris, 1934.

[32] Kohlschütter, *Z. Elektrochem.* **33** (1927) 272; Kohlschütter and Good, *ibid.* **33** (1927) 277; Kohlschütter and Jakober, *ibid.* **33** (1927) 290; Kohlschütter and Uebersax, *ibid.* **30** (1924) 72.

[33] Kohlschütter and Torricelli, *Z. Elektrochem.* **38** (1932) 213.

[34] Fischer, *Z. Elektrochem.* **49** (1943) 343, 376; **55** (1951) 92.

[35] Dahms, Green, and Weber, *Nature* **196** (1962) 1310.

[36] Fischer, *Z. Electrochem.* **54** (1950) 459.

[37] Brandes, *Z. Physik. Chem. Leipzig A* **142** (1929) 97.

[38] Mehl and Bockris, *J. Chem. Phys.* **27** (1957) 818.

[39] Parsons, *Trans. Faraday Soc.* **47** (1951) 1332.

[40] Lorenz, *Z. Naturforsch* **9a** (1954) 716.

[41] Gerischer, *Z. Elektrochem.* **62** (1958) 256.

[42] Fleischmann and Thirsk, *Electrochim. Acta* **2** (1960) 22.

[43] Kaischew and Mutaftschiew, *Z. Physik. Chem. Leipzig* **204** (1955) 334.

[44] Burton, Cabrera, and Frank, *Phil. Trans. Roy. Soc. London, Ser. A* **123** (1951) 299.

[45] Vermilyea, *J. Chem. Phys.* **25** (1956) 1254.

[46] Steinberg, *Nature* **170** (1952) 1119.

[47] Barton and Bockris, *Proc. Roy. Soc. London, Ser. A* **268** (1962) 485.

[48] Bunn and Emmet, *Discussions Faraday Soc.* **5** (1949) 119.

[49] Lighthill and Whitham, *Proc. Roy. Soc. London, Ser. A* **229** (1955) 281.

[50] Price, Vermilyea, and Webb, *Acta Met.* **6** (1958) 524.

[51] Bockris and Damjanovic, *Modern Aspects of Electrochemistry*, Vol. 3, Edited by Bockris, Chapter 4, Butterworths, London, 1964.

[52] Conway and Bockris, *Proc. Roy. Soc. London, Ser. A* **248** (1958) 394.

[53] Conway and Bockris, *Electrochim. Acta* **3** (1961) 340.

[54] Mott and Watts-Tobin, *Electrochim. Acta* **4** (1961) 79.

[55] Fischer, *Electrochim. Acta* **2** (1960) 50.

[56] A. K. Reddy, *J. Electroanal. Chem.* **6** (1963) 141.

[57] Pangarov, *Electrochim. Acta* **7** (1962) 139.

[58] Vermilyea, *J. Chem. Phys.* **27** (1957) 819.

[59] Vetter, *Elektrochemische Kinetik*, Springer-Verlag, Berlin, 1961.

[60] Inman, Bockris, and Blomgren, *J. Electroanal. Chem.* **2** (1961) 506.

[61] Mattson and Bockris, *Trans. Faraday Soc.* **55** (1959) 1586.

[62] Despic and Bockris, *J. Chem. Phys.* **32** (1960) 389.

[63] Bockris, Drazic, and Despic, *Electrochim. Acta* **4** (1961) 325.

[64] Bockris and Kita, *J. Electrochem. Soc.* **108** (1961) 676.

[65] Bockris and Enyo, *J. Electrochem. Soc.* **109** (1962) 48.

[66] Bockris and Drazic, *Electrochim. Acta* **7** (1962) 293.

[67] Bockris and Enyo, *Trans. Faraday Soc.* **58** (1962) 1187.

[68] Bockris and Kita, *J. Electrochem. Soc.* **109** (1962) 928.

[69] Fleischmann and Thirsk, *Advances in Electrochemistry*, Vol. 3., Edited by Delahay and Tobias, Chapter 3, Interscience Publishers Inc., New York, 1963.

[70] Lorenz, *Z. Physik. Chem. Frankfurt* **17** (1958) 136.

[71] Lorenz, *Z. Physik. Chem. Frankfurt* **19** (1959) 337.

[72] Lorenz and Möckel, *Z. Elektrochem.* **60** (1956) 507.

[73] Bockris and Conway, *J. Chem. Phys.* **28** (1958) 707.

[74] Bump and Remick, *J. Electrochem. Soc.* **111** (1964) 981.

[75] Nomarski and Weill, *Rev. Met.* **52** (1955) 121.

[76] Damjanovic, Paunovic, and Bockris, *Plating* **50** (1963) 735.

[77] Barret, *Am. Inst. Mining Met. Engrs., Inst. Met. Div., Tech. Pub. No.* 1865 (1945).

[78] Reddy and Bockris, *Proceedings of the Symposium on Ellipsometry in the Measurement of Surfaces and Their Films*, Edited by Passaglia, Stromberg, and Kruger, p. 229, National Bureau of Standards, Washington, 1963.

[79] Reddy, Devanathan, and Bockris, *Proc. Roy. Soc. London, Ser. A* **279** (1964) 327.

[80] Damjanovic, Paunovic, and Bockris, *J. Electroanal. Chem.* **9** (1965) 93.

[81] Hagyard and Williams, *Trans. Faraday Soc.* **57** (1961) 2288.

[82] Burbank and Wales, *J. Electrochem. Soc.* **111** (1964) 1002.

[83] Farnsworth, Schlier, George, and Burger, *J. Appl. Phys.* **29** (1958) 1150.

[84] Blomgren and Bockris, *Nature* **186** (1960) 305.

[85] Swinkels, Green, and Bockris, *Rev. Sci. Instr.* **33** (1962) 18.

[86] Stoebe, Hammad, and Rudee, *Electrochim. Acta* **9** (1964) 925.

[87] Bowden, *J. Colloid Sci.* **11** (1956) 555.

[88] Gurney, *Proc. Roy. Soc. London, Ser. A* **134** (1931) 137.

[89] Butler, *Proc. Roy. Soc. London, Ser. A* **157** (1936) 423.

[90] Marcus, *Can. J. Chem.* **37** (1959) 155.

[91] Marcus, *Transactions of the Symposium on Electrode Processes*, Edited by Yeager, p. 239, John Wiley & Sons, New York, 1961.

[92] Sacher and Laidler, *Modern Aspects of Electrochemistry*, Vol. 3., Edited by Bockris and Conway, Chapter 1, Butterworths, London, 1964.

[93] Parsons and Bockris, *Trans. Faraday Soc.* **47** (1951) 914.

[94] Matthews, Thesis, University of Pennsylvania (1964); Bockris and Matthews, *Proc. Roy. Soc. London, Ser. A* **292** (1966) 479.

[95] Bockris, *Quart. Rev.* **3** (1949) 173.

[96] Andersen and Bockris, *Electrochim. Acta* **9** (1964) 347.

[97] Azzam, *Z. Elektrochem.* **58** (1954) 889.

[98] Bernal and Fowler, *J. Chem. Phys.* **1** (1933) 515.

99 Nekrassov, private communication.
100 Hurlen, *Electrochim. Acta* **7** (1962) 653.
101 Hurlen and Lunde, *Electrochim. Acta* **8** (1963) 741.
102 Bockris and Potter, *J. Chem. Phys.* **20** (1952) 614.
103 Bockris, Devanathan, and Müller, *Proc. Roy. Soc. London, Ser. A* **274** (1963) 55.
104 Lorenz, *Z. Elektrochem.* **57** (1953) 382.
105 Gerischer and Tischer, *Z. Elektrochem.* **61** (1957) 1159.
106 Damjanovic and Bockris, *J. Electrochem. Soc.* **110** (1963) 1035.
107 Kita, Enyo, and Bockris, *Can. J. Chem.* **39** (1961) 1670.
108 Barker, *Transactions of the Symposium on Electrode Processes*, Edited by Yeager, p. 325, John Wiley & Sons, New York, 1961.
109 Smits, *The Theory of Allotropy*, Longman's, London, 1922.
110 Samoilov, *Discussions Faraday Soc.* **24** (1957) 141.
111 Bockris, Piersma, Gileadi, and Cahan, *J. Electroanal. Chem.* **7** (1964) 487.
112 Blomgren, Bockris, and Jesch, *J. Phys. Chem.* **65** (1961) 2000.
113 Buckley, *Crystal Growth*, John Wiley & Sons, New York, 1951.
114 Cottrell, *Theoretical Structural Metallurgy*, p. 174, St. Martin's Press, New York, 1959.
115 Rosenberg and Winegard, *Acta Met.* **2** (1954) 342.
116 Ling Yang, Chien-yeh Chien, and Hudson, *J. Electrochem. Soc.* **106** (1959) 632.
117 Hamilton, *Electrochim. Acta* **8** (1963) 731.
118 Seiter and Fischer, *Z. Elektrochem.* **63** (1959) 249.
119 Seiter, Fischer, and Albert, *Electrochim. Acta* **2** (1960) 97.
120 Nabarro and Jackson, *Growth and Perfection of Crystals*, Edited by Doremus, Roberts, and Turnbull, p. 13, John Wiley & Sons, New York, 1958.
121 Wranglén, *Electrochim. Acta* **2** (1960) 113.
122 Parsons, *Modern Aspects of Electrochemistry*, Vol. 1., Edited by Bockris, Chapter 3, Butterworths, London, 1954.
123 Economou, Fischer, and Trivich, *Electrochim. Acta* **2** (1960) 207.
124 Green, Swinkels, and Bockris, *Rev. Sci. Instr.* **33** (1962) 18.
125 Bockris and Swinkels, *J. Electrochem. Soc.* **111** (1964) 736.
126 Sato, *J. Electrochem. Soc.* **106** (1959) 206.
127 Damjanovic, Setty, and Bockris, *J. Electrochem. Soc.* **113** (1966) 129.
128 Turner and Johnson, *J. Electrochem. Soc.* **109** (1962) 798.
129 Pick, Storey, and Vaughan, *Electrochim. Acta* **2** (1960) 165.
130 Barnes, Storey, and Pick, *Electrochim. Acta* **2** (1960) 195.
131 Hillson, *Trans. Faraday Soc.* **50** (1954) 385.
132 Howes, *Proc. Phys. Soc. London* **74** (1959) 616.
133 Volk and Fischer, *Electrochim. Acta* **5** (1961) 112.
134 Grahame and Soderberg, *J. Chem. Phys.* **22** (1954) 449.
135 Grahame, *J. Amer. Chem. Soc.* **80** (1958) 4201.
136 Grahame and Parsons, *J. Amer. Chem. Soc.* **83** (1961) 1291.

[137] Wroblowa, Kovac, and Bockris, *Trans. Faraday Soc.* **61** (1965) 1523.

[138] Blomgren and Bockris, *J. Phys. Chem.* **63** (1959) 1475.

[139] Schapink, Oudeman, Lev, and Helle, *Trans. Faraday Soc.* **56** (1960) 415.

[140] Frumkin, *Z. Physik* **35** (1926) 792; *Ergeb. Exakt. Naturw.* **7** (1928) 235.

[141] Wroblowa and Green, *Electrochim. Acta* **8** (1963) 679.

[142] Aniansson, *J. Phys. Chem.* **55** (1951) 1286.

[143] Bockris, Green, and Swinkels, *J. Electrochem. Soc.* **111** (1964) 743.

[144] Dahms and Green, *J. Electrochem. Soc.* **110** (1963) 1075.

[145] Langmuir and Schaefer, *J. Amer. Chem. Soc.* **59** (1937) 2400.

[146] Delahay and Trachtenberg, *J. Amer. Chem. Soc.* **79** (1957) 2355.

[147] Vermilyea, *J. Electrochem. Soc.* **106** (1959) 66.

[148] Bentley and Humphreys, *Snow Crystals*, McGraw-Hill Book Co., New York, 1931.

[149] Kardos and Foulke, *Advances in Electrochemistry*, Vol. 2, Edited by Delahay and Tobias, Chapter 4, Interscience Publishers Inc., New York, 1962.

[150] Rogers and Taylor, *Electrochim. Acta* **8** (1963) 887.

[151] Watson and Edwards, *Trans. Inst. Metal Finishing* **34** (1957) 167.

[152] Weil and Paquin, *J. Electrochem. Soc.* **107** (1960) 87.

[153] Blum and Meyer, *Modern Electroplating*, Edited by Gray, Chapter 1, John Wiley & Sons, New York, 1953.

[154] Denise and Leidheiser, *J. Electrochem. Soc.* **100** (1953) 490; Roth and Leidheiser, *J. Electrochem. Soc.* **100** (1953) 553.

[155] Clark and Simonsen, *J. Electrochem. Soc.* **98** (1951) 110.

[156] Leidheiser and Gwathmey, *J. Electrochem. Soc.* **98** (1951) 225.

[157] Hoar, *Trans. Inst. Metal Finishing* **29** (1953) 302; **39** (1962) 166.

[158] Matthews, Mutucumarana, and Wilman, *Acta Cryst.* **14** (1961) 636.

[159] Ke, Hoekstra, Sison, and Trivich, *J. Electrochem. Soc.* **106** (1959) 382.

[160] See, for example, Moelwyn-Hughes, *Physical Chemistry*, p. 45, Pergamon Press, New York, 1957.

[161] Wagner, *J. Electrochem. Soc.* **98** (1951) 116; *Plating* **48** (1961) 997.

[162] Balashova, *Z. Physik. Chem. Leipzig* **207** (1957) 340.

[163] Graf and Weser, *Electrochim. Acta* **2** (1960) 145.

[164] T. B. Reddy, *J. Electrochem. Soc.* **113** (1966) 117.

[165] Kasper, *Trans. Electrochem. Soc.* **77** (1940) 353, 365; **78** (1940) 131.

[166] Sundarajan, Rajagopalan, and Reddy, *Electrochim. Acta* **8** (1963) 831.

[167] Kardos, *Proc. Am. Electroplaters' Soc.* **43** (1956) 181.

[168] Foulke and Kardos, *Proc. Am. Electroplaters' Soc.* **43** (1956) 172.

[169] Damjanovic, Paunovic, Setty, and Bockris, *Acta Met.* **13** (1965) 1092.

[170] Mullins and Hirth, *J. Phys. Chem. Solids* **24** (1963) 1391.

[171] Hulett and Young, *J. Phys. Chem. Solids* **26** (1965) 1287.

[172] Hulett and Young, *J. Electrochem. Soc.* **113** (1966) 410.

[173] Beacom and Riley, *J. Electrochem. Soc.* **106** (1959) 309; **107** (1960) 785; **108** (1961) 758.

[174] Kruglikov, Kudriavtsev, Antonov, and Dribinski, *Trans. Inst. Metal Finishing* **41** (1964) 133; Kruglikov, Kudriavtsev, Vorobiova, and Antonov, *Electrochim. Acta* **10** (1965) 253.

[175] Edwards, *Trans. Inst. Metal Finishing* **39** (1962) 33, 45, 52; **41** (1964) 140, 147, 157, 169.

[176] Raub, Baba, Knödler, and Stalzer, *Trans. Inst. Metal Finishing* **41** (1964) 96.

[177] Gerovich and Rybalchenko, *Zh. Fiz. Khim.* **32** (1958) 109.

[178] Cabrera and Burton, *Discussions Faraday Soc.* **5** (1949) 40.

[179] Frank, *Discussions Faraday Soc.* **5** (1949) 48.

[180] Budewski, Bostanoff, Vitanoff, Stoinoff, Kotzewa, and Kaischew, *Phys. Stat. Sol.* **13** (1966) 577; *idem, Electrochem. Acta* **11** (1966) 1697.

[181] Bertocci, *J. Electrochem. Soc.* **113** (1966) 604.

[182] Epelboin, Froment, and Lestrade, *C. R. Acad. Sci., Paris* **258** (1964) 4738.

Index